# YOUR HORSE
*A Step-by-Step Guide to Horse Ownership*

by Judy Chapple

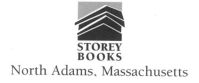

STOREY
BOOKS

North Adams, Massachusetts

*The mission of Storey Publishing is to serve our customers
by publishing practical information that encourages personal independence
in harmony with the environment.*

---

*This book is dedicated in loving memory
to my father, Donald Metz, who got me
started; to Joan McHale and Sally
Cunningham, who saw to it that it was
finished; and to my dear son Matt, who
supported me every step of the way.*

---

*Book design by Andrea Gray*

*Illustrations by Cathy Baker*

*Cover photo by Margaret Thomas*

*Inside photos by Ross Chapple unless noted otherwise*

The information in this book is true and complete to the best of our
knowledge. All recommendations are made without guarantee on the
part of the author or Storey Publishing. The author and publisher dis-
claim any liability in connection with the use of this information. For
additional information, please contact Storey Books, 210 MASS
MoCA Way, North Adams, MA 01247.

Storey Books are available for special premium and promotional uses
and for customized editions. For further information, please call
Storey's Custom Publishing Department at 1-800-793-9396.

*Printed in the United States by Versa Press*
*25  24  23  22*

**Library of Congress Cataloging-in-Publication Data**

Chapple, Judy, 1943-
     Your horse.

     Includes index.
     1. Horses.   I. Title
SF285.C25    1984         636.1         84-22280
ISBN 0-88266-358-5
ISBN 0-88266-353-4 (pbk.)

# Contents

Photo by Margaret Thomas

# Acknowledgments

◄•►

I have lots of people to thank for helping with this book. Thanks to the horsemen and vets who put up with my endless queries, to friends and neighbors who polished up their tack and horses for photographs, and to Marilee Carpenter, who can type and spell even at 3 a.m. Most of all, I thank my editors, Fred Stetson and Jill Mason, who rode the trail patiently with me, encouraging me all the way. It's been fun, it's been challenging, and it's been downright difficult at times, but the time has been well spent. I hope to have a positive influence on the lives of a few young horsemen, and I pray that this book can serve as a guide for those who want to share the adventure.

# Preface

In the story of *The Black Stallion*, a young boy, shipwrecked on a desert island, tames a wild stallion and becomes his friend for life. In countless westerns, we have seen cowboys perform all kinds of miraculous and heroic feats with the help of their faithful steeds. American culture is rich with horse stories, and we all want to believe in these beautiful tales—but these tales are fantasies.

Like any relationship, the horse-owner relationship requires a lot of hard work, care, attention, sensitivity, and respect. Sometimes the romantic notions that beginning horse owners have about horses prevent them from exercising good judgment. One common example involves the parents who buy a young, inexperienced horse for their child, so the two of them can "learn together." The horse becomes unmanageable, the child ends up terrified of horses, and the parents are out a considerable sum of money. As an instructor-trainer involved primarily with family-type horses, I have often seen the disastrous consequences of our unreal expectations.

The goal of this book is to give potential or beginning horse owners a realistic idea of what they will have to invest of themselves in the horse-human relationship and what they can expect in return. I seek to separate the myth from reality.

Photo by Margaret Thomas

# Choosing a Horse

I was born in the country and grew up on a farm in Pennsylvania. My first memory, at age three, is of "riding" my pony. I was seated in a wicker basket that my father had rigged to the pony by means of some leather straps. My legs were so short that the real saddle was reserved for practice riding on the back of the living room couch.

A few years later our family acquired a forgiving Pinto named Geronimo, who, with enduring patience, allowed Mom, Dad, and four kids to make endless mistakes on him. The only vengeance he ever took, when severely peeved, was to remove our hats and drop them in the water bucket.

Afternoons after school, that horse would be waiting when we walked up the lane. He knew he had time to race to the farthest fence line while we dropped off our books, but, gentleman that he was, he would meet us at the barn to be tacked up, resigned to the chore of carrying a bunch of kids.

Geronimo was a big horse, about 16.2 hands, and he must have wondered about the four Lilliputian kids standing on boxes to clean him. One toss of his head would have spilled his miniature grooms all over the aisle. If it hadn't been for him, we never

would have outgrown our ignorance and taken the next step.

It was Geronimo who took us to our first little backyard shows. We were so proud of him we wanted to show him off to everyone. Our confidence bloomed along with our new-found ability. Now, as I look through photograph albums, I marvel that he never took advantage of our inadequacies. He remained calm and pleasant always. Though we loved him, we never knew how rare he really was. He was a teacher and a statesman in every way.

Later, more suitably mounted on ponies, we created wonderful Indian-style hide-and-seek games in the tall corn fields, only a piece of bailing twine wrapped around the pony's jaw for a bridle. We'd gallop, the August sun burning, up and down the corn rows—big, solid yellow corn ears slapping our bodies black and blue. We didn't mind. Then there were gymkhanas, lectures, films, clinics, Pony Club, fox hunting, and horse shows. And so my childhood produced a bank of warm memories and provided a beginning that has led me through 30 years of involvement with horses.

Riding is a wonderful opportunity for

◄Choosing a horse to buy for the first time, or for the twentieth, can be a thrilling and fascinating experience.

people of all ages. Unlike most other athletes, a rider can improve his effectiveness well into his fifties. Horsemanship is a multifarious sport containing a broad range of possibilities and specialties. There's an adventure in store, tailor-made, for everyone.

To a child, whose days are crowded with authoritative adults, a pony is a means of freedom. Children are told when to get up, when to go to school, when to do chores and homework, and when to go to bed. With a pony, for a few hours at least, the child is boss. With just a tug on a lead rope, the pony follows the child, stands to be tacked up, and goes obediently off to work or play. When kids need to talk, their most private and urgent concerns, whispered to their beloved ponies, are met with calm and patient understanding. To a child, the intense active loving of a pony is returned tenfold.

And there are other advantages. The physical work of riding a horse or pony and managing a barn is good for young bodies. Children look forward to the responsibilities attached to owning their own ponies. They learn much about hard work, sportsmanship, patience, winning, losing, and economics. Above all, they have an opportunity to learn and improve forever. (And to adults who are still young at heart, exactly the same things apply.)

What I intend to discuss in the following pages is aimed at helping you select a suitable horse and then set up a sensible system of care and management to keep him healthy and sound. Although this book is for the beginner, I intend the information to be broad-based enough to provide insight to those with a longer exposure to horses. I am not encouraging the reader to think of the backyard horse as a financial investment to turn into a profit. Buying and selling horses is an extremely high-risk business which is best left to the professionals.

## THE HUNTER

The horses to which I will be referring are called the hunter type. A hunter type is the sort of horse who moves with long, low strides and carries his head in front of him, enabling him to see the ground. A horse who moves in this fashion has the ability to go through rough terrain for long distances with speed and economy of motion. A hunter type has natural ability as a jumper. There is no special breed called a hunter; he is simply a type. But some breeds that fall into the hunter category are the Quarter Horse, Morgan, Thoroughbred, Appaloosa, and Arabian. Some breeds of ponies of the hunter type are the Shetland, Welsh, Connemara, and Ponies of the Americas.

Since so many people are going back to the land and working it with horses as well, it seems appropriate to mention the draft horse. The draft horse, the big horse, is bred for the purpose of pulling or driving. He is a heavy-bodied, big-boned animal known for being sturdy and good natured, patient and strong. Percherons (the most popular draft breed in the United States), Belgians, and Shires are able to withstand the strenuous, tedious, heavy work involved in logging and plowing. When Percherons are crossed with the hunter type, they gain agility and are used as riding horses—usually big, handsome, and sound. Clydesdales (they pull the Budweiser wagon) are generally seen as teams in the show ring at horse shows.

►One of the most well-known draft horses, the Clydesdale often leads a more glamorous life than others of his type.

Breeding and conformation (the physical structure of a horse) are closely related, and they are explicitly responsible for what a horse is capable of doing. If you are interested in trail riding, showing (hunter or western division), endurance riding, cow punching, barrel racing, jumping, three-day events, dressage, or just saddling up for a full-moon, all-night ride, then you probably already own or will wind up buying a hunter type. If your interest is in heavy work such as pulling or plowing, the draft horse will probably become an important ally.

## A Proper Match

One of the most fascinating challenges I have found during my years with horses is matching the proper horse to the rider. The result, like a good marriage, is not easily achieved.

Poor combinations can have bad results; for instance, a timid rider on a bold horse is a dangerous combination. Good combinations, such as a tense horse being put to ease by a relaxed rider, are the key to success.

Buying a horse should be approached cautiously and patiently. The process of finding a suitable horse may take a few weeks or several months. One reason it is time consuming is that you'll want your horse to be versatile, and many horses aren't. A good trail horse may be terrible in the show ring. A horse who can plow a garden may trip going cross country. A pony who will pull a cart may be "cold backed" (a horse who bucks when mounted or when the girth is tightened). A horse who has been standing around in a field by himself all his life may go berserk when stimulated by a group of horses. Some horses, perfectly quiet on the trail, are afraid of barking dogs, or won't cross a hard-topped road.

Horses, like people, have their idiosyncracies, likes and dislikes. If this is a first horse, your best protection is to buy one who has been successful as somebody else's first horse. No one should be offended by the question, "What has this horse done?" Whenever possible, check it out.

There are a number of places to find horses that might suit you. A little friendly competition is good for the soul, and horse owners often get into various kinds of competitive events in order to be around people with similar interests and improve their skills by watching and talking with others, including the judges. Horse shows, rodeos, trail rides, local clubs—these events are good places to look for horses; they provide ready-made opportunities to watch horses perform. Other sources are breeding farms, riding schools, and private owners and dealers. Many people advertise in the paper or in various horse-oriented magazines such as *The Chronicle of the Horse*, published weekly in Middleburg, Virginia.

## HORSE SALES

Another source is the sale, which, for our purposes, has two categories. The Interstate Horse Sale usually offers only horses, of a pretty decent but perhaps mixed quality. The livestock sale, on the other hand, offers swine, sheep, and cattle, as well as horses. It is held in most rural counties on a monthly or bimonthly basis. Your agricultural extension agent or local newspaper can give you the location. The sale is the great smorgasbord of horses—a feast for the eyes, a colorful, noisy, confusing, fast-moving event that will delight you one moment and tug at your heart the next. I've sat in the

bleachers many times, feet propped up on the seat in front of me, to watch the broadest possible range of horse flesh parade through the ring. There are some horses who are older than Moses, some too young to be weaned, some so quiet you could preach a sermon from their rumps. Some are crazier than hell and some are great athletes. There are fat horses, thin horses, sound horses, horses who are broke, horses who aren't broke, and horses who are in some way broken down. In 1981, two horses competing as jumpers on the international level were sold at stock sales.

Although you might find any kind of horse you want at the sale, world economics have played a devastating hand to the horse industry in general and specifically to the availability of grade horses (non-specific breeding). About 40,000 horses are exported each year from the East Coast to France and Canada for meat. When horse meat is selling for fifty cents a pound, someone will pull an unused pleasure horse out of the field and convert him into a new couch, a car payment, or a trip to Disney World. Many of the good old souls, the real "packers," horses who have taught whole families to ride, are winding up on ships. It has made the grade horse, a tried and true family friend, harder to find and more expensive.

If you go to a sale, go early. First, find the sale manager and tell him what you are looking for. Ask him if he knows any owners or dealers who are bringing such a horse. Sometimes he may be able to help you. Some of the larger Interstate sales (which offer catalogues to help you make selections) employ a vet for the day. Although he may not be an equine specialist, he can give the horse you select a quick examination. He will usually be able to spend from five to ten minutes with you to examine the horse for general soundness of heart, lungs, and eyes. He will discuss major conformation faults, if any, with you. He can analyze whether the horse appears to be healthy and moves well. But he can't really be sure if the horse is sound without x-rays, and he won't have time or facilities to do them. A standard equine veterinarian examination on a private farm or at a dealer's stables would last from a half hour to two hours, depending on the kind of work (physical stress) that a horse would be put to. What you get at a sale is better than nothing, but it isn't thorough. And remember, at the local stock sales, even this service isn't usually available.

## Drugs

I encourage you to have your selection thoroughly vetted (examined by a vet), because, sad to say, as drugs have become an epidemic in our society, they also have invaded the entire horse industry. Drugging horses is an extremely sophisticated science and a common practice. As soon as a quick and reliable test comes along, someone figures out how to mask the test with another drug for which there is no test. It is not difficult to acquire the drugs. An unethical dealer could have a horse on three legs this morning and sell him as sound in the afternoon with the help of 10cc of phenylbutazone. Or, if a horse is a little hot or crazy, perhaps 2cc of acepromazine will produce the well-mannered, beautifully schooled horse of your dreams—for about three hours.

How can you know if the horse who blinks too much, whose eyes seem slightly glazed, is drugged or just sleepy? How can

◄ Potential buyers listen carefully as a young horse is put up for bid during a typical livestock sale.

you tell if the horse with a little bump on his jugular vein has been stuck by a needle or by a horse fly? Without a blood or urine sample, you really can't.

Drugs are definitely more prevalent in higher-priced horses, simply because there is more money at stake. If I were at a stock sale and saw a well-bred, big, attractive horse, perhaps with his mane braided, I would approach buying that horse extremely cautiously; there's a good chance that he's got a problem. Big, good-looking horses should command good prices—more than the stock sale could bring. Some years ago, a friend of mine bought a big, handsome Thoroughbred at a stock sale. The next day the horse began to sweat and stayed on his feet by propping himself against his stall walls. It turned out that he was having severe withdrawal symptoms. Buying a horse at a stock sale is risky.

## THE BEST SOURCE

The dealer, agent, friend, or private owner is the source I recommend, for three reasons:

1. The buyer may study the horse's performance and ride the horse more than once. He isn't rushed into making a decision.

2. The buyer may question the owner about the horse.

3. The buyer may have a vet declare whether the horse is sound and what his chances are of remaining sound. That is to say, a vet can give his opinion as to whether the horse is suitable for the work he will do when you buy him.

Dealers, agents, or private owners may be found by word of mouth, by inquiry at a large horse farm, through riding instructors, in the local paper, or through a veterinarian. You may never find a source that everybody agrees is good, because of the subjective nature of the topic. Do the best you can, bearing in mind that the vet's advice will be invaluable at the time you make your selection.

Folks who sell horses usually have a variety of them moving through their barns all the time, and they spend many hours on the road looking at horses who are just coming on the market. Many dealers are, or have been, professional riders. They have an uncanny memory for detail and a keen sensibility about how to analyze a horse. And they analyze a customer just as readily. The best way to deal with them is to be completely straightforward. Tell them exactly what you are looking for and how you intend to use the horse.

"My daughter wants a backyard horse she can take around to the little shows. She's 5"6' and has been riding once a week for a year." If this is your first horse and you have never ridden anything but a bicycle, tell him so. Describe to him the kind of facilities that you have at home. Tell him how much pasture you have, whether you have a run-in shed or a barn, and whether the horse will be in the company of other horses or alone.

In considering the type of horse to select, it may be helpful to take into account the types of horses that are popular in your area. The advantages are numerous. It's fun to meet people with similar interests, and there will be clubs and organizations already set up that you can join.

Whatever you decide to do, there are two organizations that are excellent for kids and that teach the fundamentals of horse care, stable management, and riding. The first is the United States Pony Club, which involves kids up to the age of 18. It offers nothing but equestrian activities. Those who have gone through the ranks of the Pony Club have a very solid background, and some emerge as potential top riders. The other, of course, is the 4-H, which has several very good programs, one of which is horse oriented. To find out where the nearest chapters are, call or write:

The United States Pony Club
893 South Matlack Street
Suite 110
West Chester, PA 19382

National 4-H Council
7100 Connecticut Avenue
Chevy Chase, MD 20815
(*or contact your local county extension office*)

## Sex

The word "colt" is a widely misunderstood term. It does not mean a young horse of either sex. Folks say, "My mare is going to have a colt." Actually, the mare will produce a colt or a filly; and until it is known which, it's referred to as a *foal*. A *colt* is an unaltered male until he is three, at which time he becomes known as a *stallion*. A stallion is generally used in this country for racing or breeding. Because of his aggressive temperament, he is unsuitable as a family horse. An altered male is called a *gelding*. A *filly* is a female until her third year, at which time she is referred to as a *mare*.

A *dam* is a mother; a foal is "out of" a certain dam. A *sire* is a father; a foal is "by" a certain stallion. The foals produced by the sire are his "get." One might say, "The sire got 16 foals last year. All of his get are chestnuts, out of Thoroughbred dams."

Either sex has its advantages. Since geldings are altered (or "cut") they are not subject to stimulation by a mare's monthly metabolic changes, or "heats." Although mares may be bred, they are susceptible to female-type infections, and some get fussy when they are in heat. It can be embarrassing to try to ride a mare with love on her mind. If you want a backyard horse, the sex is purely a matter of preference. But if you are considering competing, a gelding could give you the winning edge.

## Age

There exists such widespread and unfair prejudice toward the older horse that he is often referred to as "aged" or "smooth-mouthed" after only his 12th year. Some people say seven is the perfect age. Perfect for what? Although horses have keen memories, they are very slow and methodical learners. The harder the task, the longer it takes. Seven is too old for a galloping horse, but a field hunter, dressage horse, or polo pony usually is not at his best until he is at least ten. Horses, like people, show age at different times. Idle Dice, a celebrity horse on the show jumping circuit, was still winning at 18.

Family horses, under much less mental and physical stress, may go on longer. I've seen many people buy a first horse who is 15 or 16 years old and have several years of good, safe fun. By the time the horse is ready to be retired, the owner is advanced enough to go to something more challenging. Horses who are sound during their early years tend to stay sound. If a horse is suitable for you, if he knows how to do what

**Without valid registration papers, one way to estimate a horse's age is by examining his teeth. The following illustrations show what to look for.**

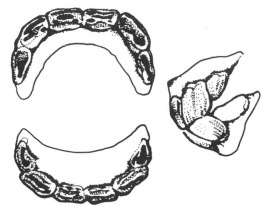

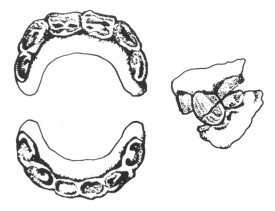

**At 1 year:** All temporary incisors are visible. Centrals and intermediates are in contact, and the surfaces of the centrals show wear. Upper and lower corner incisors do not contact.

**At 2½ years:** Permanent central incisors are coming in, but are not yet in contact. The lower permanent teeth may not be entirely free of the gum.

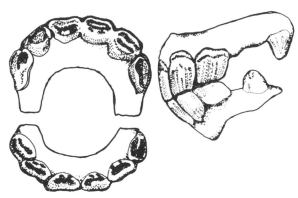

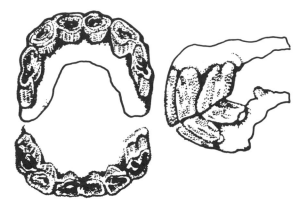

**At 5 years:** All permanent teeth are in. The centrals and intermediates show wear on the chewing surface, but cups are still visible and are completely encircled with enamel.

**At 10 years:** Teeth are increasingly slanted from the jaw line. Chewing surfaces of the lower centrals and intermediates are rounded. Upper intermediates are nearly smooth.

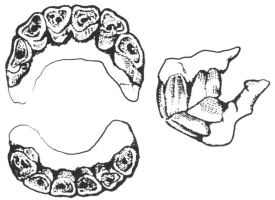

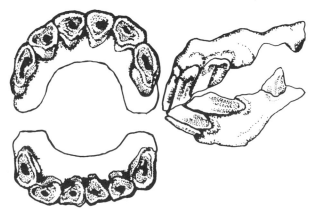

**At 15 years:** Lower incisors appear shorter than uppers from the front. A groove ("Galvayne's groove") appears on the upper corner incisor. The chewing surfaces of the lower centrals and intermediates appear triangular.

**At 20 years:** Slant of the teeth is quite oblique, and Galvayne's groove extends the length of the corner incisor. Incisors are triangular. Teeth have space between them, and the lowers may be worn almost to the gum line.

Source: *How to Buy the Right Horse* by Margaret Cabell Self

you want him to do, don't let his age stop you. You may have found the perfect match.

On the younger end of the spectrum, consider five as border line if this is your first or second horse. By six, a horse has generally settled somewhere near his potential. That is to say, he has defined what his attitudes toward working are. A two-, three-, or four-year-old may be forgiven for some of his mistakes and bad manners, but if he's still making the same mistakes at six, forget him. A novice rider should not attempt to correct a bad mannered horse.

## Size

The size of a horse is measured in "hands." A hand is a 4-inch span (about the distance across the average man's palm). A pony is 14½ hands (14.2) or smaller. Anything larger than that is a horse. Thus, it's possible to have an animal bred as a horse—a Thoroughbred, let's say—which, because he never reached full size, is a pony. Horses are measured from the withers to the ground with a (measuring) "stick."

The size of the horse must be appropriate for the rider. A good basic principle of

A simple measuring stick will indicate how many "hands" a horse measures from his withers to the ground.

measurement is that when the rider is mounted, with stirrups properly adjusted, his foot should be about even with the bottom of the girth (see diagram of horse, p. 38). Combined with his balance, a rider's leg is his stickum, his glue. As the rider improves, his leg is used in a multitude of ways, a sophisticated language that encourages the horse to move forward, to bend into a circle, to slow down, and even to back up.

Since controlling the horse requires effective use of the rider's calves by putting pressure against the horse, it is of utmost importance that this principle of leg placement be understood. A rider too big for his horse will have part of his lower leg swinging in the breeze, rather than on the horse. A rider too small for his horse will have too much calf against the saddle—again, an ineffective state of affairs. A rider who is tall from the waist up needs a horse with a long neck, and a very heavy rider should have a short-backed horse.

But the size of the horse does not tell all. Someone who looks fine on a narrow horse who is 15.2, for example, could also be proportionally correct on a very big-bodied horse of 15.0 hands. The broader the horse, the more leg he takes up.

If you are very tall and need a really big horse, consider 16.2 as tall enough and try to find a horse big-bodied enough to take up your leg. Horses who are much taller than 16.2 generally don't have the added muscle and tendon strength needed in proportion to their size to support their increased weight. They tend to have more problems staying sound, as well as increased incidences of respiratory ailments, than smaller horses.

Now a word about ponies. If you are

looking for a pony for your family, your job may be a little harder. Ponies, some will say, are nasty and try to take advantage of a child. Others say that ponies, like cedar trees, aren't any good until they're 35 years old. It just ain't so. There is nothing intrinsically wrong with ponies as breeds. The problem with many ponies is that most adults are too big to ride them, and consequently they are broken by kids, a lot of whom are novice riders. So the pony learns that he may do whatever he pleases.

Good ponies, properly broken, are available and are worth their weight in gold. Their advantages are that they are easy keepers, are hardy, and tend to stay sound. They fit the child and are much more easily tacked, untacked, and cleaned. Some require front shoes, some no shoes at all, just trimming. And there is the added attraction that a spill from a pony, who is close to the ground, is far less traumatic than one from a horse. If a child is accidentally kicked or stepped on by a pony (especially one without shoes), chances are that nothing will hurt but his pride. So if the size of your child requires a pony, which is likely, take time, be patient, and a good one will turn up sooner or later.

## PREVENTING INJURY

Before going to see and ride your first horse, you must realize that the most uncertain and potentially dangerous part of the entire process is when a novice mounts an unknown horse. Horse dealers may have had stock recently shipped in, and some of those horses may have quirks of which even the dealer is unaware. Although the great majority of falls are merely minor incon-

veniences, there are several sensible precautionary steps you can take to protect yourself just in case.

It's a good idea to take a trip to your local tack store to buy a hard hat or helmet. There are several kinds to choose from, but since hats tend to come off during a fall, a chin strap is essential. I know very well how the argument goes—they're too expensive ($25-$55)—but the point is that if you can't afford a helmet, you can't afford a horse. When your children object, explain to them that professional riders always wear their hats. They are just as essential a part of the gear as the saddle is. Do yourself a favor, buy the hat.

Another problem that comes up all the time is kids riding in sneakers. Because they have no heel, sneakers will easily slip through the stirrup. If that happens, the rider can't fall free, away from the horse. A smooth-sole leather shoe or boot with a heel will give needed protection and comfort. Some folks insist on going into a barn in sandals. A real quick cure for this habit comes about when the sandaled foot finds itself under the shod hoof. Cracked toes or bruises and a tetanus shot are usually all it takes to get the idea.

## THE BIG DAY

So, after considering it and discussing it with the family, you've finally agreed that you are going to do it. You've decided that today's the big day—you're going to start to look for a horse! You load the kids into the pickup and drive down to the general store. The local newspaper has a sale section. Everyone gathers around while you rip open the pages. There! You see it! "FOR SALE pleasure horse 15.1 h. gray m. 10 y.o." The mare is in the barn of a dealer who has been recommended to you. Let's go to look at this fascinating and mysterious creature, the horse, whose spirit is so entwined with man's.

11

# A Little History

That horse you love, the gray out in the barn, has only recently come home. He was gone for millions of years. His name was Eohippus, the Dawn Horse, and he began his career in North America and Europe about 60 million years ago. His fossils have been found in Wyoming, Utah, Colorado, and New Mexico, as well as England and Europe. Eohippus was small as a fox and had four-toed feet in front and three-toed feet in back. He was a browser and very shy. His striped, yellow, camouflaged body helped him to hide among the forest leaves, and his padded toes enabled him to move quietly. But he was able to survive the hostile environment primarily because he was equipped with very fast reflexes, excellent vision, and speed.

As the Ice Age came and went, geological changes resulted in a change of Eohippus's source of food. Eventually, the Great Plains stretched across what had once been forest and swamps, and the horse no longer dined on soft leaves but chewed tough pampas grasses. He developed sharp incisors to clip off the grass and molars to grind it, all encased in a deep jaw bone that allowed his teeth to grow continuously so that he

couldn't wear them off. And as his jaw changed, he also grew.

As he roamed, seeking the good grasses, his pasture land led him to the Bering Strait. He crossed over the land link that then existed and moved to Europe and Asia. Rising waters from melting ice caps eventually covered this land passage, and he was gone from North America for a long, long time. Those who remained behind died out, perhaps as a result of bacterial infection.

By the time man swung down from his tree, lost his tail, and began walking in an upright position, Eohippus had evolved from fox-size to pony-size and his feet had only one toe, now a hoof. Two of his non-functional toes had withdrawn to become splint bones, still a part of his skeletal structure. The hoof enabled him to run still faster. And run he did, for man was after him.

Even before man learned to write, he painted the walls of the Lascaux caves in southwestern France, depicting ponies fleeing from showers of arrows, being eaten, and being used for clothing. The cracked bones of about 10,000 horses around the caves indicate that Stone Age man found

◄The horse's association with man has taken many forms—their stories have been linked from the beginning of time.

1 3

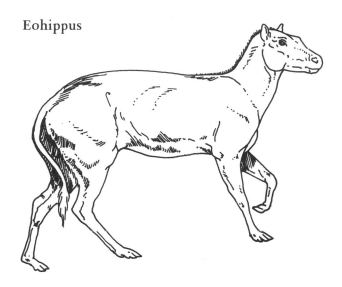

Eohippus

him to be delicious. (That was about 20,000 years ago, but in some areas of the modern world the palate of man is still treated to the sweet taste of horse meat.)

## FOUR TYPES

During this period, the various climatic and geographical conditions in Europe and Asia encouraged the horse to develop into four distinct types. In the mountainous, damp, and rocky coastal area of western Europe, the Celtic pony roamed and developed a water-resistant coat which helped protect him from the chilled, wind-whipped rains. From this rugged creature a great many pony breeds have developed—all tough-footed and hearty. One cousin grew tall on the lush forest grasses of western Europe and was cultivated to become a war horse for knights in armor.

A second type of horse developed on the frigid Mongolian Steppes and in China. He became resistant to cold. His slitted, ram-like nostrils breathed in only so much of the freezing air per breath, and his large internal air passages warmed it before it reached his lungs. In the late 19th century, a Russian explorer discovered a herd of these pre-historic horses living in Mongolia. Named for him, they are known as Prewalski's Horse. They are yellow, like Eohippus, and retain a dorsal stripe. They exactly resemble drawings and paintings sketched on bone fragments and stones during the Stone Age.

The third type settled in central and southern Asia and was the largest of the primitive-type horses. He became a war horse, eventually entered North Africa, and then migrated to Spain via a long-submerged land bridge. He is the Barb, forefather of the Andalusian breed of Spain.

The fourth type of horse is referred to as the Caspian pony. He lived in Mesopotamia, a great and early center of agriculture, and in nearby hot and arid regions. He developed heat resistance and his wide, flaring nostrils drew in lots of air to fill his blood with oxygen as he fled from his predators over huge distances across barren plains. This horse became the present-day Arabian.

## MAN'S HELPER

It is not known just how it happened; perhaps horses were kept live for a food supply and man got to know them. But eventually, as man became a cultivator, the horse became his helper. There is no question that the history of the world would have been very different had it not been for man's friend, companion, and dependable ally, the horse.

For centuries, all over the East, skilled nomadic horsemen challenged one another's tribal and grazing lands, continually exchanging customs in addition to real estate. The destruction of Rome at the beginning of the Dark Ages was accomplished with the help of the horse. Attila the Hun swept, pillaged, and plundered on top of the horse. As warfare techniques changed, the horse did too.

At first, cavalry soldiers held their weapons, engaging in hand-to-hand combat. Their horses were small and agile, the better to pivot and run. During the Crusades in the 11th century, knights wore chain mail and used lances as weapons. They needed bigger horses, but agility was still necessary.

As the invention of the longbow changed warfare again, selective breeding took on another new dimension. Soldiers and their war horses wore extremely heavy plate armor to protect them from the terrific onslaught of arrows that would otherwise

A winch was sometimes used to lift knights onto their horses because their armor was so heavy and cumbersome.

have accounted for their mass slaughter. The plate armor was so weighty that men were set on top of their horses with pulley ropes. If man or horse fell, he lay there, often unable to get up again. These war horses were brave, patient, and massive. They are the forefathers of our draft horses.

As history progressed and fire power was introduced, horses made such good targets that they were often withdrawn from battle and used in harness, pulling guns and supply wagons.

## AMERICA

When Columbus came to America in 1492, there were no horses here. It was the Spanish conquistadors in the early 16th century who brought their spotted horses (with acute herding instinct from their experience in the bull ring) back to America. Some escaped and migrated north to the western plains, where they contributed to the instinctual cow sense and herding ability of the Quarter Horse. These horses were called mustangs, and those that the Indians didn't capture roamed wild over the plains in huge herds. Wild herds lured the white man's valued horses away from his farm and ate the sparse grasses needed for cattle. Thus, the mustang lost favor with the white man, who prized him most when converted into fertilizer. The battle still rages. Those who seek his demise fight bitterly with his preservationists. It seems odd that, perhaps for the first time since the Stone Age, man has no use for these little horses.

Though the wild mustang became unpopular with man, the horse remained an important ally. It was largely horse power that created this country. Horses cleared our fields, moved our cattle, dragged our battle guns, carried our mail, pulled our barges and trolleys, and provided sport which has evolved into a huge industry. We have used him for work and pleasure and financial benefit.

## The Horse Industry Today

The following facts, based upon data compiled by the American Horse Council, give some idea of the nature and size of the horse industry in the United States:

- Approximately 3.2 million horse owners keep an estimated 8.3 million horses, 80 percent of which are owned for recreational use. (The remaining 20 percent are used for a variety of profit-oriented activities, including agriculture, logging, racing, and professional exhibition.)
- Since 1960, annual foal registration by 16 major horse and pony registries has grown from 72,853 to 260,973, an increase of 358 percent.
- More than 250,000 young people are engaged in 4-H horse projects, an enrollment larger than the number of cattle and swine projects combined.
- Horse racing has been America's number one spectator sport for over 30 years, and all horse sports combined draw more than 100 million spectators each year.

Jousting is a popular horse sport. Instead of lancing a competitor, the rider must lance the ring—while riding at a dead gallop.

- Horse sports generate more than $1 billion in revenue for federal, state, and local governments each year, not including property and employee income taxes.
- According to a survey by the United States Department of the Interior, 27 million people ride at least once a year, more than half of them on a regular basis.

Throughout history people have held few things dearer than their horses. Some have risked their necks to steal them. Arabs brought them into their tents. Richard III offered his kingdom for one. They are loved for their beauty and mystery as well as for the status they convey. But perhaps the most reasonable explanation for their enduring high regard is their remarkable athletic ability, which human beings have harnessed to complete what we could never do alone. The horse is a keen athlete for whom the rider is the mind, and their combined talents lead to endless possibilities.

# *Breeds*

—◄•►—

A breed is a homogeneous group of animals that have common characteristics which are inherited by succeeding generations. By selective breeding, man has made more changes in the horse over a few centuries than nature did in millions of years of evolution. We have been very clever in getting the utmost from him. From the need for speed in the English cavalry the Thoroughbred was developed. To fulfill our requirements to work cattle we bred the Quarter Horse. Plantation owners who rode fence lines all day checking the fields developed the American Saddlehorse and the Tennessee Walker. Those who sought financial fulfillment at the trotting track produced the Standardbred. We have been extremely sophisticated in breeding for our explicit purposes.

Yet to say, for example, that the Quarter Horse is kind and has cow sense, or that the Thoroughbred is high-strung and very fast, is to speak about the breed type as an ideal. Individuals within breeds differ tremendously. There are good and bad horses within all breeds, and we must look carefully at each horse we are considering for selection.

Purebred and Thoroughbred horses are usually going to have bigger price tags than half-bred horses, and are generally selected for a job that requires specializing, that uses the outstanding talents that the breed has developed. However, specialties require many years of hard work for a rider to achieve satisfactory results on a competitive level. Few kids who start out with the idea that they want to be a jumper rider, for instance, have enough experience to make that decision, and often become discouraged or distracted and leave the sport. I believe that it's most sensible to begin with a nice, well-mannered horse who can do a little of everything. Most likely he won't be brilliant at any of it, but neither will the rider, to begin with.

Select for the desired characteristics of the breed and realize that some of these characteristics may be found in half-bred horses as well. Half-breed registries record that a certain horse, out of an unregistered mare is by a registered stallion. It may be useful to have papers on a registered mare (for future breeding purposes), but it's not relevant for geldings. Many families make good selections by choosing half-bred or "cold-

◄ A Connemara Pony and his young rider demonstrate excellent form in one of the talents the Connemara is known for.

blooded" horses (horses who have some draft-horse blood).

To give you an idea what characteristics are found in various breeds, I've listed some breeds in this chapter that seem to be most successful in fulfilling the role of family horse. I've chosen these breeds because they have natural tendencies, due to their attitudes and the way they move, to be able to perform (at some level) both on the flat and over fences.

# COLOR

Before discussing the breeds, it would be useful to know a little about color and how it relates to breeds. Coat color is influenced by the horse's type of pigment and its structure and distribution throughout the hair. While a pretty coat is nice to look at, the coat's practical value is that it helps regulate body temperature. Like thermal underwear, it traps a layer of air next to the skin.

We know Eohippus was yellow with a dorsal stripe, perhaps not beautiful, but mother nature's very own sleight-of-hand trick—now you see him, now you don't. He simply blended into the earth's colors—a useful stunt in a hostile environment. As the horse became domesticated and thereby protected, two things happened: he no longer needed his camouflage; and man cultivated color mutations and bred for colors that pleased him.

In selecting their first horse, some buyers are tempted to buy for color alone. However, the long-term relationship with a horse takes on a complexity in which the horse's soundness, temperament, and ability form

---

► Stockings are white markings on a horse's legs that go up as far as the hock or the knee; socks go above the ankle.

the base; color doesn't tell us much more about the horse than his color.

Breeders are fussy about color. Thoroughbreds, Arabians, Morgans, Quarter Horses, and most other major breeds accept only solid color into their registries. (Solid color includes white markings and black points.) Some breeders register their stock primarily around color or color patterns, admitting individuals of various breeds that simply have the proper color requirements. The Appaloosa, Palomino, Pinto, and Paint are color breeds.

*White markings* are areas of white hair on the horse's face and on his legs, below the knee and hock. Unless white markings extend higher up the legs or to the sides of the face, they have no relationship to the markings of a Pinto or Paint.

*Points* refer to the legs, mane, and tail. A bay has black points. He is brown with a black mane and tail, and his legs are black

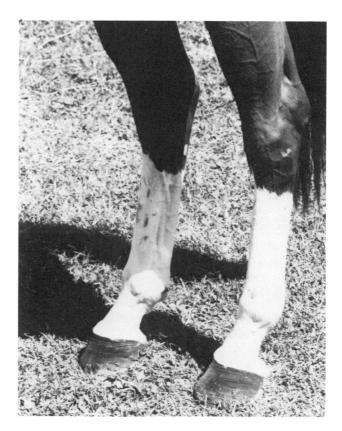

up to his knees and hocks. But a bay may also have white markings—for example, a blaze and two hind socks.

## White

Some people confuse gray horses and white horses, but they are really easy to distinguish. A white horse has pink skin and a pink muzzle. He is born with a white coat. White horses are uncommon.

## Gray

A gray horse is born with a dark or black coat that lightens gradually with maturity, but he always has dark skin and muzzle. With age, some grays get freckles or dapples, and sometimes their coats lighten to white.

## Buckskin

A buckskin has a sooty yellow coat, dorsal stripe, and black points.

Photo by Margaret Thomas

## Palomino

A Palomino has a coat from "the color of a gold coin" to several tones lighter or darker. His mane and tail are white, silver, or cream, with no more than 15 percent dark hairs mixed in.

## Chestnut or Sorrel

Chestnuts or sorrels are reddish-brown horses of varying tones. They have the same color points as their coat color.

## Bay

A bay has a red, auburn, or brown body with black points. These are blood bays, sandy bays, and dark bays, respectively.

## Brown

A very dark brown horse—almost black— or even black, with a brown muzzle.

## Black

A horse with a black coat and a black muzzle, an uncommon color.

## Roan

A roan has a base coat of color with an even sprinkling of white hairs throughout. A roaned chestnut is a strawberry roan. A roaned black is a blue roan. A roaned bay is a red roan.

## Piebald

A piebald horse is colored only with black and white patches.

## Skewbald

A skewbald horse is colored with patches of black, brown, and white.

◄A blaze is one example of white markings which can be found on a horse's face and on his lower legs.

## THE ARABIAN

The Arabian is the oldest known breed. Because of his long history of domestication and association with man, he is a horse of kind and willing temperament, intelligence, and versatility. There are many individuals of the breed who make good family horses if great size is not required.

The Arabian was scarce and costly in the United States until quite recently. Before the end of the Second World War there were only about 3,000 horses registered in the Arabian stud book. But after the war, the United States Army imported a number of Polish-bred Arabians who had been captured from the Germans (who took them from Poland), and their number has increased dramatically since 1960. Their blood has figured more prominently in the upgrading of other breeds than any other. Three Arabian stallions, the Godolphin Arabian, the Byerly Turk, and the Darley Arabian, were imported to England in the 1700s and became the foundation sires of the Thoroughbred.

The odd fact that the Arabs prized their mares more highly than their stallions made the importation of the Arabian stud horse more practical and less expensive, and was responsible, along with economics and climate, for a decline in the number of Arabians in their native land and an increase elsewhere.

Because of his small size and late arrival in large numbers on the American scene, the Arabian missed out being known as a specializer. By the time the Arabian began competing in the United States, the Quarter Horse had the corner on the working-horse and short-distance-racing markets. The Thoroughbred was unexcelled as a steeplechaser, hunter, jumper, and distance horse, and the Saddlebred was winning as a gaited horse. Though the Arabian is versatile, since 1950 he has earned his reputation as an endurance horse, winning 100-mile trailride competitions. His talent and ability probably have much to do with his early days as a desert war horse, traveling alongside camels over great distances with little food or water. He is tough, has hard bone, and tends to stay sound. Generally, he has one less lumbar vertebra than other horses, accounting for his short, strong back; and he has two fewer bones in his tail. When crossed with the Thoroughbred, he is called an Anglo-Arabian.

## Conformation and Characteristics:

Short head from poll to muzzle; flaring nostrils; very prominent wide-set eyes; somewhat undefined withers; short back with high, flat croup; high-set, flaring tail; and rather large feet. He is small, 14.0-15.1 h., and averages about 900 pounds. His mane, tail, and coat are exceptionally fine and glossy. He is strong and has an even temperament.

Arabian Horse Registry of America, Inc.
12000 Zuni Street
Westminster, CO 80234

# THE THOROUGHBRED

Generally speaking, it will be more difficult to select a suitable family horse from the Thoroughbred breed than from the other breeds I am describing. I am including him because he is so popular, so numerous, and so talented and famous that it is impossible to disregard him. He is a glamour horse, and he takes himself very seriously. Historically, breeders have selected sires and dams pretty much on the basis of speed rather than conformation, and breeders have produced a wonderful and efficient galloping machine. But because we begin their training when they're so young, many Thoroughbred horses develop soundness problems.

The Thoroughbred began in England in the 17th century. Three imported Arabians were the foundation sires of the Thoroughbred line, and all Thoroughbreds can be traced to one of those three stallions. In 1730, Bull Rock, a son of the Darley Arabian, arrived in Virginia to stand stud. He was 21 years old. By 1745 Thoroughbred breeding and racing had begun to take hold in this country, but it wasn't until 1863 that racetracks really began to cater to the Thoroughbred breeding industry.

Because the Thoroughbred has such a wide range of types, he excels at a broad range of athletics. Other breeds can do all the things that the Thoroughbred can do—it's just that the Thoroughbred does them better. He is a sprinter and a distance horse. He is a dressage horse, a polo pony, a show jumper, a steeplechaser, and a field hunter. He is competitive, keen, and high-strung.

◀ The Arabian's early experience as a desert war horse probably contributed to the strength and stamina of the breed today.

Thoroughbred blood has had great influence on other breeds. At one time the Army Remount Service stood some 1,000 stallions in various parts of the country. Their breeding services were instrumental in upgrading and refining the foals of cold-blooded and half-bred mares. Thoroughbred blood runs heavily in the veins of the American Quarter Horse, the Standardbred, the Morgan, the American Saddlehorse, and the Appaloosa.

## Conformation and Characteristics:

Great variety. Usually a refined head, long neck, prominent withers, and depth through the girth. They range in size from 15.0 to 17.0 h. and are of all solid colors, with or without white on legs and face. They are competitive and courageous.

A family who likes the Thoroughbred type but hasn't found the right horse may be successful with a Thoroughbred-Quarter Horse cross.

The Jockey Club Thoroughbred Registry
40 East 52nd Street
New York, NY 10022

## THE QUARTER HORSE

Most people think of the Quarter Horse as a western horse, but he began his career in the East. He was assigned his name in early Colonial times because of his specialty. Before racing was really accepted, English and Irish settlers—hard-core, church-going men—punched trails through the forest so they could secretly race their horses and engage in a little friendly, backyard wagering. (They must have been dedicated to

---

► Perhaps best known for his extraordinary herding ability, the Quarter Horse excels in many other skills as well.

24

broad-saw trails wide and straight enough for several galloping horses.) These men satisfied themselves that a quarter mile, or a "quarter path," as they called it, offered a perfect test for speed: the simple, hushed beginning of a raging, complex industry.

Many of the early foundation sires were bull-dog types. They were heavily muscled, mutton-withered, short-legged, and powerfully built for short bursts of speed. But this breeding isn't as popular today. The original foundation sire is thought to be Janus, a grandson of the great Godolphin Arabian, and there has since been a generous infusion of Thoroughbred blood into the Quarter Horse line.

When the settlers moved west, pushing their cattle and Quarter Horses, they bred their horses to the mustang, the horse brought here by the Spanish conquistadors in the 16th century. Since the Spanish were great bullfighting enthusiasts, their horses had a good herding instinct, which, when mixed with the characteristics of the Quarter Horse, developed into "cow sense," an instinctive ability to herd cattle. Given his speed and cow sense, the Quarter Horse became a first-rate cutting horse who can "stop on a dime and give you a nickle change."

It's common to see Quarter Horses competing successfully as pleasure horses and show jumpers. They are good at all types of ranch work, can endure the polo field, and are sensible and athletic field hunters. The Quarter Horse is the central focus of a huge racing industry. His versatility has increased his popularity, and he has become America's sweetheart. The American Quarter Horse Association has more registered horses than any other breed association, numbering over one million.

Many Quarter Horses are able to fulfill family tasks as well as to carry sophisticated riders in a variety of specialties.

## Conformation and Characteristics:

Shorter head from poll to muzzle than the Thoroughbred; round, wide jowl; slightly straighter shoulder and pasterns than the Thoroughbred; heavily muscled arms and gaskins; short, dense cannon bone. Unfortunately, some halter judges are looking for a very small foot to which many breeders are complying. The Quarter Horse is sure-footed, tractable, tough, and intelligent.

American Quarter Horse Association
P.O. Box 200
Amarillo, TX 79168

# THE MORGAN

Where legends are involved and few records exist, some stories are hard to trace. But it seems that somewhere in Vermont around 1790, a bay colt was foaled and named Figure. At maturity he was a compact, powerfully built, "big-little horse," with a lot of style; his lineage is uncertain, but he probably had Arab breeding. He was given to a schoolmaster named Justin Morgan, as payment of a debt. Some time later, the stallion himself came to take the name of his owner. Figure, or Justin Morgan, proved to be a willing, kind, and versatile horse, and his owner rented him to neighboring farmers to clear fields, pull stumps and logs, and to stand at stud. But he gained a repu-

tation for himself especially as a trotter. His son, Sherman, who much resembled his sire, stood at stud as well, and both had the genetic ability to throw their conformation no matter what type of mares were brought to them (which is called prepotency).

The realization dawned slowly, around the mid-1800s that there seemed to be a large number of similar-looking, dark, compact, and stylish horses who were courageous, hard working, and versatile, and very fast trotters. The best of these went to the track to compete against the top trotters of all, the Standardbreds. But since the Standardbred had the corner on the trotting market, Morgan breeders decided to infuse the Morgan's blood with Standardbred blood, as well as the blood of speed horses such as the Thoroughbred and the Quarter Horse. The effect of this inbreeding was simply to diffuse the Morgan blood, while Standardbreds continued to win at the track. Shortly thereafter, Morgan show breeders coveted the position of the American Saddlehorse and again diluted Morgan blood. The crosses were not advantageous to the Morgan, and the compact, stylish, all-around horse was hardly recognizable.

It was through the efforts of the United States Army Remount Service, as well as some private breeders who understood the value of the Morgan's traits and stood good stallions, that the Morgan resembling Figure was brought back. But that is not the end of the saga. At present, Morgan show breeders again have the notion that Morgans should emulate the American Saddlehorse and are subjecting them to the same show requirements as the American Saddlehorse. This is unfortunate, but in all

probability the breed will endure in spite of the present trend.

Outside the show ring, the Morgan is found working cattle, playing polo, at rodeos, on trails, and in the hunt field. Many individuals in the Morgan breed make good family horses.

## Conformation and Characteristics:

The Morgan is sure-footed, kind, and an easy keeper. He has a fine head and a cresty neck which he holds higher than the Quarter Horse. On the whole, he is similar to the Quarter Horse but generally has a bigger front. He is good at a wide range of athletics.

American Morgan Horse Association, Inc.
P.O. Box 960
Shelburne, VT 05482

# THE APPALOOSA

The Appaloosa may be traced back 2,000 years to Persia and China through art history. He reached American shores with the Spanish conquistadors and wandered to the western plains. In modern history, the Nez Perce Indians improved the breed by weeding out horses who lacked the heart or endurance needed to gallop after buffalo or into battle. The Appaloosa has developed good feet and bone, as well as staying power and courage. The Indians kept their herds in Utah, Washington, and Oregon, and in the cool, lush valley of the Palouse River, from which the Appaloosa got his name.

The Appaloosa has gained refinement from infusion of blood from Quarter Horses, Arabians, and Thoroughbreds. Working cattle on the range or jumping in the show ring, from west to east, Appaloosas

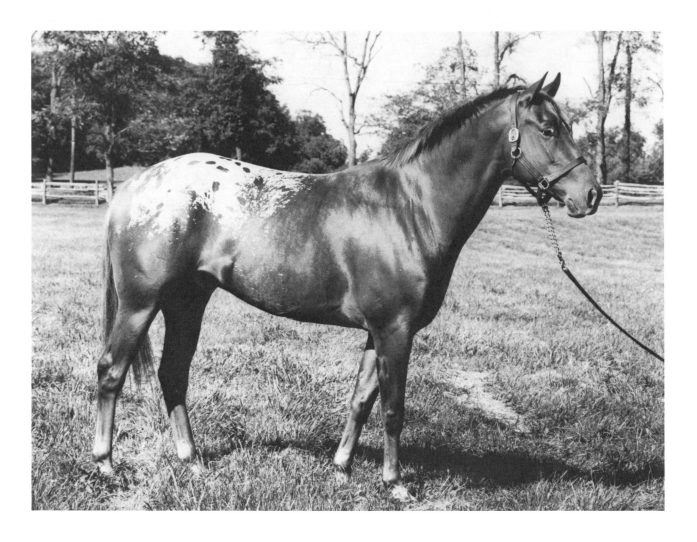

can be seen in all areas of equestrian competition. From among their ranks one is sure to find a horse suitable to carry the family in a variety of endeavors.

## Conformation and Characteristics:

White ring around the iris of the eye; parti-colored muzzle; and vertically striped hooves. Some have thin mane and tail. Common color patterns are leopard (white with black spots), solid colors with white blanket, and roans.

Appaloosa Horse Club, Inc.
P.O. Box 8403
Moscow, ID 83843

## PINTOS AND PAINTS

Pintos and Paints are spotted horses, but they have different registries. Unlike the Appaloosa, who often acquires his color gradually, the Pinto and Paint are born with a permanent pattern.

The Pinto Horse Association, formed in 1956, registers horses, as well as ponies over 12 hands, three- and five-gaited horses, pleasure horses, and others whose conformation is a good representation of their breed type. There are more types of Pintos than Paints.

The American Paint Horse Association began its registry in 1965, and its goal is to

---

►Even though color pattern determines a Paint horse, its registry is working to establish conformation as well.

◄ The versatility of the Appaloosa may be a result of his rather exotic history. His distinctive markings make him easy to recognize.

create a breed of horse over 14 hands with the conformation of a stock horse type. Registry requires a Paint to have a sire registered as a Thoroughbred, a Quarter Horse, or a Paint. It is interesting to note, however, that the American Quarter Horse Association refuses to register a Paint.

There are two basic color patterns:

*Tobiano* is a clean-cut pattern with white as the base color, mane and tail the color of the skin at its source. Often the horse has white stockings and usually brown eyes.

*Overo* has a dark base color with white spots of irregular or roan edges. Mane and tail are often roan, and blue or white eyes, known as china or glass eyes, are common. The legs are usually dark.

Traditionally, the spotted horse has been used as a parade horse or a backyard horse, but the recent registry of Pintos and Paints has refined and upgraded the breeds to the extent that he is gaining acceptance in all areas of the horse world. Many individuals from either breed would well suit the requirements of a family.

Pinto Horse Association of America, Inc.
1900 Samuels Avenue
Fort Worth, TX 76102-1141


American Paint Horse Association
P.O. Box 961023
Fort Worth, TX 76161-0023

Photo by Mary Nokes, courtesy of *Paint Horse Journal*

►Although it may take some time to find one, a Shetland Pony with a pleasant disposition can make an excellent kids' pony.

## THE PALOMINO

The name Palomino describes color rather than type; horses of various breeds who clearly represent their own breed type are accepted by the Palomino registries as long as they have the necessary color. The first registry was begun by the Palomino Horse Association (PHA) in the United States in 1936, and accepts horses with pink or yellow skin. The second registry, the Palomino Horse Breeders of America (PHBA), formed shortly after the PHA, will accept a horse with dark skin whose sire or dam is listed with the PHBA or another registry. Its members are, therefore, more elite. Both registries require horses to be between 14 h. and 17 h. and have a golden coat. Actually, the shades vary from light cream to dark gold. The mane and tail must be cream, white, or silver and contain no more than 15 percent chestnut or black hairs. White markings on the face and below the knee are permitted. A great number of Palominos are also registered as Quarter Horses and American Saddlehorses.

The Palomino horse is thought to be of Spanish extraction. Queen Isabella's soldiers were mounted on Palominos. Spanish breeders must have known something about genetics, because it's tricky to get the color. Crossing one Palomino with another produces a Palomino only about 50 percent of the time. The most reliable combination to achieve the color is a Palomino-sorrel cross.

A good Palomino is lovely to look at and has gained acceptance in areas where it used to be considered out of place. Palominos are seen competing in a wide variety of equestrian sports, under both Western and English tack. Because of the broad range of

Photo by Lori Brightnell, courtesy of the Palomino Horse Breeders of America

type, the family should be able to select one to fulfill its requirements.

Palomino Horse Association, Inc.
P.O. Box 324
Jefferson City, MO 65102


Palomino Horse Breeders of America
15253 East Skelly Drive
Tulsa, OK 74116-2620


## PONIES

All ponies must stand no higher at the withers than 14.2 h. (58"). There are many types for various uses, but most are English in origin and come from the Celtic pony, whose small size helped him cope with his rugged environment.

---

◄ Horses who are accepted by different registries may also be Palomino if they have the necessary golden coat.

## THE SHETLAND

The Shetland Pony survived for centuries on the Shetland Isles, about 200 miles north of Scotland, not far from the Arctic Circle— a cold, damp climate with sparse vegetation. Relative to his size he is the strongest breed; he is built like a miniature draft horse. In Britain he was used as a riding pony and a work pony, pulling coal carts deep in the mines. There are few left in England; they have taken up residence on American soil.

American breeders have altered the Shetland a lot. In the 1800s he was used as a kids' pony. He could be hitched up to the family cart more easily than a horse for short trips to Grandma's. But Americans wanted to do their errands more stylishly, so the Shetland was bred to be slightly taller and slimmer. This design had certain advantages for the short-legged child rider, who no longer had to do the splits to sit on him. His refinement also resulted in more defined withers, better to keep the saddle in place. But show breeders—going on the theory that if a little change is good, then more is better—continued altering him to the point that the show Shetland now resembles a miniature Saddlehorse with long feet, heavy shoes, and a fiery disposition. This peculiar change in his breed type makes the show Shetland impractical as a kids' pony.

There remains, however, the other American type, which is very close to the original and is a good choice. He's an easy keeper who requires so little food he can live on a snowball. His size makes him easy for a child to tack up, to clean, and to love. Having extremely tough feet, a result of his rocky island days, he may usually go barefoot, requiring only trimming. Because his constitution requires so little feed for him to thrive, lush pastures can cause him to founder, or develop laminitis (see page 84), so it may be necessary to keep him in a stall during the day in the summer.

For larger size, Shetlands are often crossed with Welsh ponies or other breeds. Shetlands generally outlive even long-lived horses and remain useful well into their twenties.

## Conformation and Characteristics:

Not more than 11.2 hands, solid colors and pintos. Heavy mane and tail with thick coat. Faces are dished and ears small. Easy keepers, hardy, and usually sound.

American Shetland Pony Club
P.O. Box 3415
Peoria, IL 61614

Photo courtesy of the Welsh Pony Society of America

## THE WELSH PONY

The Welsh Pony is the most popular pony in America because he is athletic and of useful size, with a good disposition. He is refined and "typey" (closely resembling the ideal of the breed), looking much like a little Arabian. He is seen in all areas of the equestrian world where ponies are used—pulling a pony cart, in the show ring, on the flat and over fences, in the hunt field, and for Pony Club and 4-H projects. He crosses well with Morgans, Arabians, Thoroughbreds, and Quarter Horses when larger size is needed. For a smaller pony, a Shetland-Welsh cross usually turns out well.

The breed established its registry in 1907. The American Welsh Pony stud book requires Section A Welsh Mountain Ponies to be 12.2 or under. Section B ponies, called Welsh Riding Ponies, are larger and stand 12.3 to 14 hands.

## Conformation and Characteristics:

Stands 11.0 to 14.0 hands; solid colors with white markings. No pintos. Very refined, breedy head with small ears and large eyes. Athletic, versatile, and strong, with few soundness problems.

Welsh Pony and Cob Society of America
P.O. Box 2977
Winchester, VA 22601

◄The most popular pony in America, the Welsh Pony has a pleasant disposition and makes a good pony for a child.

►Relatively new to this country, the Connemara Pony is often seen in the hunt field. He has stamina and a good disposition.

# THE PONY OF THE AMERICAS

The Pony of the Americas (POA) is a fast growing breed, gaining popularity in the United States, Canada, and South America. The foundation sire, Black Hand, was the result of a cross between a Shetland Pony and an Appaloosa mare. POAs are Appaloosa in color. No pintos are allowed into the registry, which was founded in 1954. The desired conformation type looks like a cross between a Quarter Horse and an Arabian. He has speed, versatility, and toughness; his way of moving enables him to perform almost any task. He is seen under Western and English tack.

## Conformation and Characteristics:

He must stand between 11.2 and 13.2 hands. Markings of the Appaloosa, dished face; versatile, attractive, and sound. Mane and tail thicker than the Appaloosa.

Pony of the Americas Club, Inc.
5240 Elmwood Avenue
Indianapolis, IN 46203

Photo by Barclay Livestock Photography, courtesy of Pony of the Americas Club, Inc.

The Pony of the Americas has the markings of an Appaloosa, the conformation of a Quarter Horse and an Arabian.

## THE CONNEMARA PONY

The Connemara Pony hails from Ireland and was first imported to the United States in the 1950s by famed Thoroughbred breeder George Ohstrum of Virginia. The Connemara is much respected for his talent as a superb jumper and for his staying power in the hunt field. He is strong enough to carry an adult and quiet enough to "baby sit" a larger child. He is versatile, tough, and sound. Because of his size, he could suit a one-horse family as an all-around horse if Mom and Dad aren't too tall and the kids too small.

## Conformation and Characteristics:

Stands 13.0 to 14.2 hands; compact, deep body; short cannon bone; fine mane and tail. Usually dun, brown, gray, black, and cream colors. White markings are common. No pintos.

American Connemara Pony Society
HoshieKon Farm
P.O. Box 513
Goshen, CT 06756

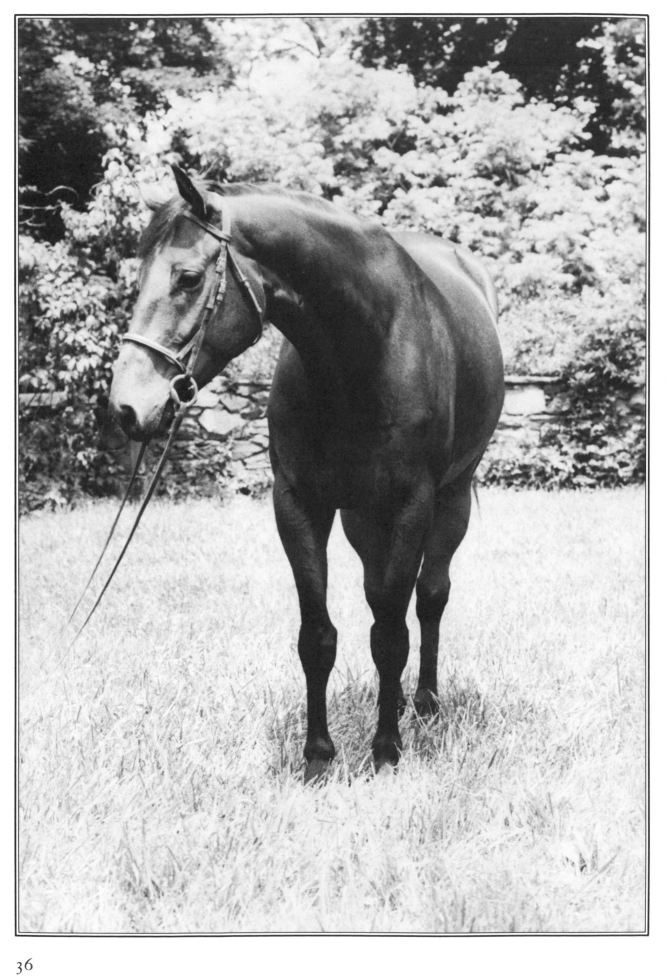

# *Conformation*

The subject of conformation is a little like the subject of religion—it takes a lot of study to really be able to analyze it. It is a fascinating subject, however, and well worth the effort. An understanding of conformation is important because if a horse's basic skeletal structure is faulty or malaligned, he will be handicapped. If you have an idea of how the horse's mechanical structure works, you will be better able to determine what your horse can and cannot do. Understanding the basic principles of conformation makes the correctly made horse much more appreciated.

You will use your horse in some way, and when you begin to think of him as a worker or an athlete (not just a pet), you will see that it is vitally important that the parts function correctly for the job. The basis on which you select your horse will include price, age, breeding, size, and sex. But most important of all are serviceability, suitability, and soundness.

Although perfect conformation is the ideal (especially for breeding purposes), there is room for some compromise without losing function. On the other hand, few horses can overcome really bad confor-

mation without having predictable soundness problems. We need to know if the horse, to put it plainly, has nature on his side. But we also need to remember that in analyzing the horse, conformation is a factor which has equal value with temperament and performance. A horse with acceptable conformation that he puts to work is more useful than a beautiful horse who hates to work. So let's kick the tires, toot the horn, and slam the door. To simplify this complex mechanical structure, we'll analyze him part by part, looking at conformation and function.

## THE FIRST LOOK

To begin with, when you go to look at a horse, have the handler present him to you on level ground in a halter or a bridle, but not with a saddle. This way, you can see his withers and back and get a good view of his top line to see how his flesh covers his spinal column. Generally (working Quarter Horses are an exception), the mane falls on the right side of a horse's neck, known as the "off" side. The left side is the "near" side, so

---

◄ A solid chest and well-spaced, straight front legs provide good support for a horse.

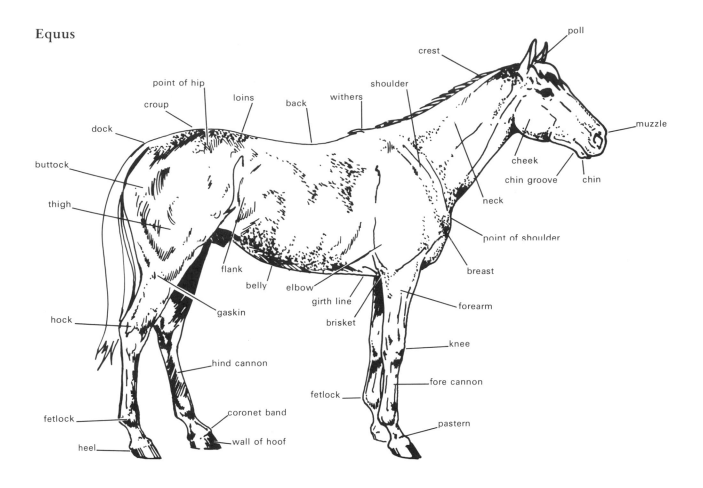

called because it is the side of the horse that is generally close to you. He is led, saddled and bridled, mounted and dismounted, all from the left.

This custom probably originated in ancient times when equestrian soldiers wore their swords on the left hip. Had they mounted from the right (or off) side, they would have had to throw their left leg and their long, heavy sword over the horse, a cumbersome and awkward matter, to say the least. Think of this near- and off-side business as standard etiquette between you and the horse, and a safety precaution as well. The horse is a slow, methodical learner, so don't startle him by changing the rules.

All this is to say, stand on the right side of him to begin with and have a look. You will notice that relative to his length he is very narrow, like a book on edge. It is easy to see from that fact alone how easily a rider's balance can affect a horse in motion. The horse's spinal column runs from just behind his ears to part way down his tail. The length of spinal column in front of the withers supports the head and neck and is very mobile and supple; whereas behind the withers the spinal column is quite rigid—a stable, structural support designed to carry

▶ This attractive horse has lots of substance and exhibits many characteristics of good conformation.

the great weight of his body. This heavy structure (the average horse weighs 900-1,200 pounds) is supported on the ground by four legs and relatively tiny hooves, which are under tremendous pressure, especially when he is required to support all his weight on one foot—when galloping or landing from a jump, for example. So, we can think of the horse as having three parts that must operate as a unit: part 1, head and neck; part 2, body; and part 3, limbs and feet.

## THE HEAD

The horse's head should be looked at carefully. It's a good gauge of whether you will rate the horse high or low. Obviously, the head is responsible for a lot of things: seeing, hearing, breathing, smelling, and, surprisingly, balance. First of all, look for signs of animation and expression. They will be reflected in his head carriage, eyes, ears, and nostrils. Like a movie star, he should have presence. He should be pleasant. A wide forehead gives more space for the brain and sets the eyes further on the outside for good vision to the side and rear. His jowls should be well-defined and rounded, spaced far enough apart for good breathing. His head should taper down to a small muzzle, the nostrils wide, ears small. Now, separate the horse's lips and look into his mouth. His gums should be a pinkish color. Very pale gums could indicate anemia. His breath should smell sweet. Foul-smelling breath could be caused by an infection of the teeth or sinus cavity, or rotting feed between his teeth and cheek, caused by poor mastication. If a horse emits

a foul smell from his nostrils, he may have a lung infection.

His upper and lower incisor teeth should meet evenly. If the lower teeth fall behind the upper teeth (overbite), he is known as a "parrot-mouthed" horse. Rather than nipping off grass with his teeth, a parrot-mouthed horse employs his tongue, using the wrap, press, and pull method, like cows. Because of this compensation, a parrot-mouthed horse can usually maintain himself quite well on long grass. But because the condition causes visible distortion (his upper lip protrudes, beak-like, over his lower lip), some people shy away from a parrot-mouthed horse; and you may want to consider that in terms of resale. While you're looking in the horse's mouth, check to see if his tongue is intact. As a child, I had a pony who was super—until he had

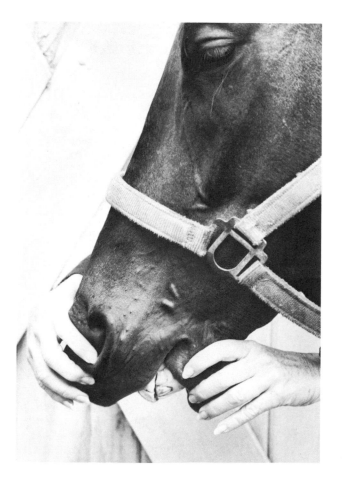

the bridle on, when he would become extremely fussy and upset. No wonder—it turned out that he had lost part of his tongue, perhaps during a fall in which it was bitten off. We solved the problem by using a hackamore (a bitless bridle) on the pony. It's a rare condition, to be sure, but it doesn't cost anything to look.

## The Eyes

Next, let's take a look at his eyes. They should be spaced wide apart and bulge out somewhat, to aid peripheral vision. A human being's close-set eyes are focused by the automatic opening and closing of the lens. The impressions that we see are transferred to the brain and interpreted as simple, three-dimensional images; this is called bifocal vision. Not so the horse.

The horse has one very narrow frontal field of bifocal vision, but the similarity stops there, for the horse's retina is shaped differently than ours, having a slightly bulging or banked surface. To account for this shape, he must put his head in the proper position to focus. When his head is raised, light strikes the upper part of the retina, and he sees objects close at hand. His head is carried in front of him to see middle distances and lowered for long distance. Except when he is using his narrow field of bifocal vision, each eye sends a separate image to the brain, and he sees two separate impressions.

I'm sure this accounts for much of horses' shying. The same old tree stump, 20 feet from the barn, may look like a scary ghost when approached from a slightly different angle. But most important, horses who travel cross country, and must be able to see where they are about to put their feet as well

◄ This horse is slightly parrot-mouthed; a more severe case can cause unattractive distortion of the horse's mouth.

When his head is raised, light rays strike the upper part of the retina to bring near objects into focus. Hence, the horse throws his head up to see objects close to his nose.

When his head is in normal standing position, light rays strike the middle of the retina, giving clear focus to objects somewhat farther away.

When his head is down, light rays strike the lower part of the retina, which is closest to the lens and gives clear focus to objects at the greatest distance.

as where they are going to be several strides away, need to carry their heads flexibly, in front of them. Very high-headed horses (due to bad conformation or bad riders) can't scan the scene properly.

While grazing, a horse can see through his legs behind him. He has excellent peripheral vision—about 340 degrees. Unless he is so fat that his body blocks his view, he can see around himself almost to his tail. That's his blind spot and, consequently, the place where many accidents occur. Horses usually

sleep standing up; they just lock their knees and nod out. Approaching a horse from behind makes it difficult to determine how alert he is. Any horse, especially a sleeping horse or a nervous horse, needs some warning that you're around. His instinct, when frightened from behind, is to kick. Don't put a hand on him before you speak to him, or he may give you a hind foot.

Get into the habit of speaking to him when you're working around him. You can be as creative about it as you want. I like to

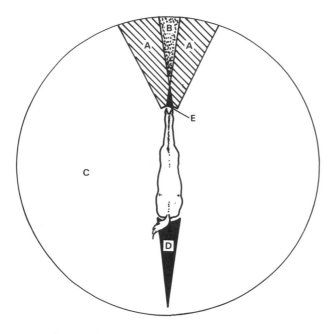

Lateral eye placement allows the horse to see either straight ahead (areas A & B) or to the sides and back (area C). He cannot see both to the front and sides without slight eye movement. There is a small area (B) that the horse can see with both eyes at the same time (binocular vision). Black areas (D & E) indicate the horse's blind spots.

sing to my horses. My family says our coon hound has a better voice, but our horses have never complained. They seem relaxed and comfortable, accustomed to my routine.

The horse has better night vision than you and I do, but coming out of a shadowy area into bright light, his eyes take longer to adjust. Science has not yet determined whether he sees in color or just some colors, but at least he sees a greater variety of black, white, and gray than human beings do.

A horse's eye should appear bold and clear. A cloudy area or a milky-looking eye could mean he has a scar or an eye disease. But even a perfectly healthy looking eye could suffer from a detached retina or paralysis of the optic nerve and be totally blind. To check whether he sees, stand by his shoulder and point a finger quickly toward his eye, causing as little air flow as possible. He should exercise his blinking reflex because he sees your finger, not because you are fanning his eyeball.

Horses who have lost an eye due to injury or infection should be judged on an individual basis. If the horse has made a good adjustment after several months, and he isn't nervous about it (and if the disability is reflected in the price), he can be a good purchase. But he would be harder to turn over if you were considering resale in the future. If you should wind up with a one-eyed horse, remember to make a special effort to let him know where you are when you're on his blind side.

## THE NECK

In thinking about the horse's structure we must always be mindful that conformation is the proportion of one part in relationship to another. To have control of the horse, a rider must control the horse's head. Since his head is the point of balance at the end of a lever that the rider controls (the neck), it's important that the head is in good proportion to the neck, and that the place where head and neck meet (the throttle) is slender, or fine. The reason is this: it's physically easier for a horse to be flexible and bend his neck at the poll while working with the bit in his mouth if his neck is not thick and unwieldy.

Horses whose work requires great balance, agility, and flexibility, such as jumpers,

►This horse features a well-formed head with an animated, kind facial expression, a long neck, and a deep, sloping shoulder.

42

dressage horses, cutting horses, and speed horses, cannot function well if they are big-headed and thick-necked. They need to be springy up front. Horses with too much up front "lay heavily into the bridle" or are "heavy on the forehand." (For the job of the draft horse, however, who never is required to lift his front end off the ground, that characteristic becomes extremely useful. He should be heavy up front so he *can* lean into his collar. His job requires more muscle than agility, and his weight is an asset. Fullness in front of the shoulder also acts as padding.)

The first two bones in the neck, the atlas and the axis, are of greater dimension than the other neck vertebrae, nature's way of protecting an area of the spinal column that is prone to injury from falls or from rearing up and hitting the head. Where the head meets the throat, there should be a wide angle to allow for good breathing space. Trace your finger down the jugular vein on either side of the neck. Too many injections can cause the vein to collapse. While there, check his pulse. It should be between 35 and 40 beats per minute. Much over or under that indicates a problem.

The neck should be long and have a slight arch or crest, a convex line running from poll to withers. Some breeds have more arch than others, and the line varies even among individuals of the same breed. Shetland Ponies and stallions tend to be very "cresty." Development of the crest is often considered in determining when to castrate a stud colt. Individuals who are getting too cresty are cut earlier to terminate a growth pattern that would cause them to look unrefined. A colt who is undeveloped in the neck is usually given more time to get a better crest before being cut.

Both of these horses tend to throw their necks upward when pressure is applied to the bit. The ewe-necked horse (A) has a light, concave neck with no flexion at the poll.

The horse with a close-coupled, "upsidedown" neck (B) cannot flex at the poll either—the heavy arch under his neck restricts collection.

Some horses are built in reverse. They have a concave line from poll to withers, and they are known as "ewe-necked." Because of this conformation fault, they often carry their head and neck too high, jaw up in the air. They are known as "stargazers," and I don't fancy the idea of galloping over a big fence on one. I'd prefer that he was flexible and could see what he was doing. A horse can't see the spot directly under his head, and when it comes time to jump, that's where the fence happens to be.

## THE WITHERS

The neck should meet a well-defined withers, the elevated portion of the spine between the neck and the back. A practical purpose of a well-defined withers is to keep the saddle in place. Horses of different breeds have more or less withers. Generally, the Thoroughbred displays a more prominent withers than the Quarter Horse, for example.

Many ponies are mutton-withered (fleshy and undefined) and need a tighter girth or cinch to keep the saddle in place. Even so, some of them are adept at the engaging practice of dropping their heads while flinging their heels (an often perfected downhill specialty), to throw the saddle forward and down the neck, child and all. To prevent this, the crupper strap was invented. It attaches to the back of the pony's saddle, and loops down under the pony's tail, where it is padded to prevent chafing, and back to the saddle again. It's a wonderful, purposeful piece of tack.

Although a prominent withers is desirable, it needs good protection from chafing

►The long neck of this lovely horse meets a well-developed withers, another important feature of good conformation.

and rubbing by the saddle. Sometimes the saddle pad or blanket isn't enough. It's important also that the saddle fit the horse properly. Two fingers' width between the saddle and the horse's spine allows ventilation and relieves pressure. If a sore should develop on the withers, it should be treated immediately and the horse rested until the sore is entirely healed.

Horses begin their courting by the gentle biting of the withers. It's an extremely sensitive area and must be kept clean and well padded. The fact is, what may begin as a small swelling or bruise on the withers could end up a disaster. If the sore progresses to infection, a condition known as fistula of the withers may develop. Since the location is at the top of the horse's spine, the infection drains down between the shoulder blades and the spine. The condition can worsen until a hole develops that a man could plunge his fist into. A cure is long-term and hard to come by. Fistula of the withers may result in the destruction of the horse.

To eliminate this possibility, be sure that the horse's back and pad are clean and dry before saddling. It's best to wash the pads after each use, but if this proves impractical, a clean towel under a semi-clean pad will do the trick.

## THE SHOULDERS AND LEGS

The withers are supported on either side by the shoulder bones (scapula), which are flat and triangular, covering the first six or seven pairs of ribs. The shoulder should slope at about a 45-degree angle. Because the

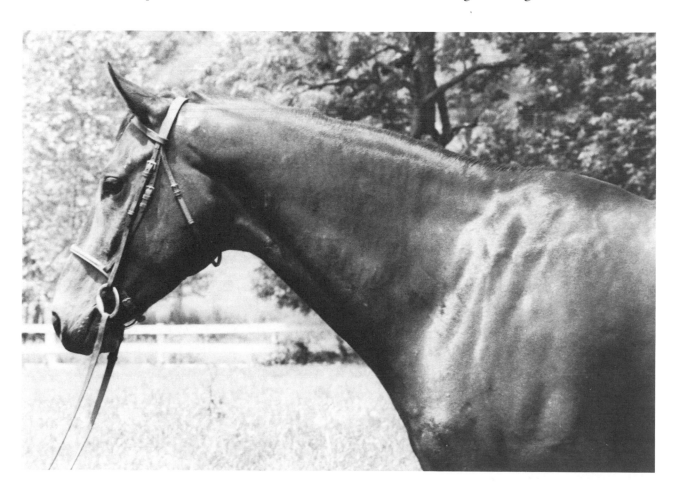

shoulder is the main point of attachment of the forearm to the body by muscles and ligaments (better shock absorbers than bone to bone), there is an intimate relationship between the function of the shoulder and the forearm. Put simply, the shoulder shapes the stride. The more sloping the shoulder, the further forward the forearm and foot can move, causing a long, smooth stride.

Conversely, a short, upright angle at the shoulder is often repeated in the pastern, resulting in a short stride that is jarring as well, because the short, straight pastern isn't well designed to absorb shock. In movement, the front end of a horse reaches, balances, and absorbs shock while the hind

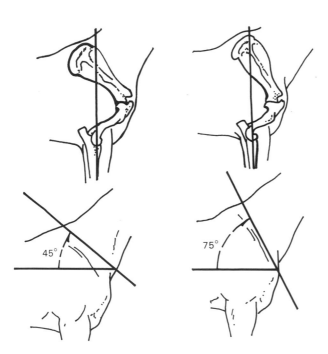

The well-angulated shoulder on the left allows smoother action than the upright shoulder on the right. As the shoulder angle approaches a straight line, stress upon the limb increases due to jolting concussion upon joints with limited flexibility.

end propels it forward. I imagine that if Eohippus had known what Equus knows, he would have hung onto his padded toes, for mother nature didn't intend the horse to do much of the athletic work required of him by man, work which puts stress and strain on his powerful physique.

To get an idea of the kind of power we're dealing with, consider the fact that a horse can shatter a foreleg during a gallop simply because his muscles are out of sync. Gravity being what it is, a horse's weight takes the most direct route to the ground. It follows that the bone linkages in a horse's forelegs must be aligned properly. Chance meetings won't do. Flat joint surfaces must meet other flat joint surfaces (between cartilage) at their strongest point, not slightly to the left, here, there, or anywhere else. Some horses, observed standing on a flat surface, will reveal legs that are not straight. Yet their movement is true. Conversely, some straight-legged horses move crooked. Given a choice of the two, take the horse who moves straight every time.

## THE FEET

Squat down and eyeball the horse from the front. His feet should be well proportioned for his size, and they should be exactly the same size and shape. The center of the foot should be directly under the center of the ankle. This imaginary line continues up the cannon bone through the center of the knee. The forearm should continue this straight line up to the point of the shoulder.

Now, let's test this alignment a bit further. Standing in front of the horse, lift his

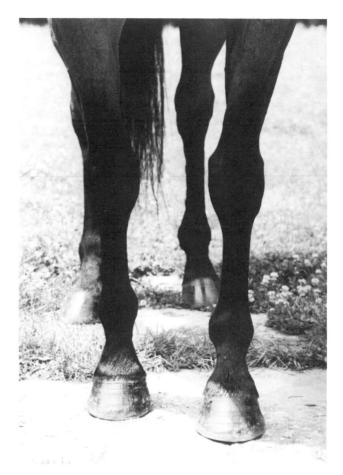

forearm just above the knee straight forward and allow the leg to swing free. Does the foot fall inward, toward the other leg, or outward, toward the breeze? We want it to fall straight down, knee and foot under the forearm, demonstrating that our straight imaginary line is still in effect. If it is, rejoice. But we've still got to watch the horse move.

I can hear a chorus of people shouting, "My horse doesn't move straight and he's perfectly sound!" That may be true, and if it is, the owners are very lucky, especially if their horses are working regularly. Nobody can guarantee anything, but a horse with good legs who moves straight has a much better chance. These principles are true relative to what kind of work the horse will do. A horse used once a week on Sundays to wander quietly through the woods may deviate from the ideal more successfully than those worked regularly. Your equine vet will offer invaluable assistance when it comes time to make the decision.

Viewed from the side, the angle of the foot should continue through the pastern joint, to enable the unit as a whole to serve as a shock absorber. Legs should not be "tied in" and should be neither "over at the knee" nor "back at the knee."

## THE HOCKS

Most horse people prefer short-backed horses. One often sees photographs of speed horses that are noticeably long in the back, but remember that they are carrying very light cargo, perched out of the saddle and over the withers. Heavy riders certainly do best by selecting short-backed horses. The back should run smoothly into well-

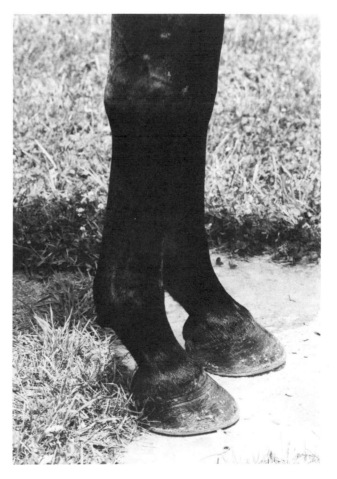

►The angle of the foot should continue up through the pastern joint, creating a good shock absorber.

47

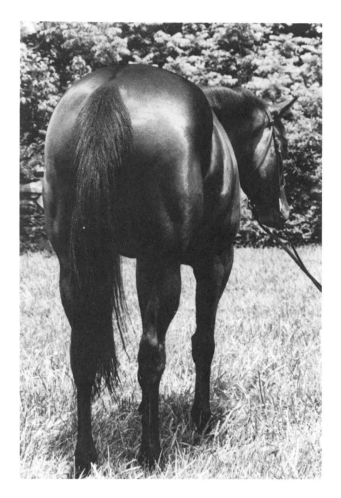

muscled and deep quarters with a good distance from point of hip to point of buttock. Now, another imaginary vertical line should fall from the point of buttock to the hock, then to the fetlock, and to the ground. A hind leg that extends beyond the vertical is said to be "camped out." Horses whose hocks stand in front of the vertical line are "sickle-hocked," and those whose hocks slope inward when viewed from behind, are said to be "cow-hocked." Any of these malalignments cause stress to the hock joint.

As the front end of a horse lifts and reaches, it is the hind end that pushes all this weight forward, and much of this impulsion comes from the hocks. Because it is engaged during all movement, the hock is the largest and hardest-worked joint in the equine body. It is composed of six bones. A horse who is used for pulling heavy loads (pushing, really, since he pushes into the collar); for jumping, where impulsion comes from the hocks; for dressage, wherein he is very precisely balanced; or for cutting cows, which requires pivoting and sliding stops, needs to have this joint in good working order. Viewed from the front, it should be broad; from the side it should be flat; and from the inside, slightly rounded. Given that impulsion comes from the hind end, the rear pastern will be more upright and the hind foot more elliptical, better for the needed leverage and ground gripping.

▶ In the extended trot of an elegant dressage horse and rider you can see the impulsion coming from the hocks.

48

## THE VET TEST

The art of selecting a good horse comes down to knowing the difference between a horse who already has or may develop a serious problem, and one who has an insignificant problem. The reason we must make these choices is that there's no such thing as a perfect horse. This is where you can use the help of a vet.

Some buyers go to great pains and expense to bring in a vet from a neighboring county on the theory that they must not employ anyone who knows the owner or works on his horses. They fear that somehow the facts might be juggled to the advantage of the owner. I have never seen this happen. Get the best vet you can, very preferably an equine vet, and don't worry about him. He represents the buyer, and he looks very inept if he passes an unsound horse.

It's pretty much a standard practice these days to x-ray the front feet, and the vet will help you decide whether the horse has any suspicious areas that a radiograph would clarify, perhaps an ankle or a hock. If you or the vet wonder whether the horse may have been given pain killers or tranquilizers, it's usually worth the cost of a drug test to ease the mind.

If the horse doesn't pass the vet, you have at least avoided being stuck with an unsound or unsaleable horse, and you have learned something more about what to look for and what can occur. If the horse does pass, congratulations! There are few thrills as exciting as bringing home a new horse. 🐎

Photo by Margaret Thomas

# *Barns*

When we start working horses daily, we diminish their ability to cope with the weather. We condition them and keep them fit, so they need more fuel for their engines; we shorten their manes, tails, forelocks, and fetlocks, which protect them from bad winter weather and summer flies; we put them in small pastures, but we often don't quite get around to a good program of parasite control; and we ship them here and there, exposing them to stress and disease. The point is that the more we require them to do, the better care we need to give them. The barn becomes a part of that better care as well as a convenience for us, a center where we can strictly adhere to a daily schedule of care, all the tools of the trade within easy reach.

Most horse barns are hubs of year-round activity where secure, relaxed horses enjoy eating and sleeping while riders hustle around cleaning tack, mucking stalls, and sweeping aisles. The barn is where the family dog waits excitedly to follow you on your ride, where cats stalk mice, and where children whisper secrets to their ponies. On many farms the barn is the culture center. It's where you meet other riders, the vet, the blacksmith, and the feed man. A barn doesn't need to be fancy, but it should be convenient, well planned, and organized.

## BUILDING A BARN

If you're building a barn, one of the first factors to consider is its distance from your front door. It should be far away enough so that the flies and the odor don't join you in your living room, but close enough so that it's not a chore to get to, no matter what the weather is. If you live in a cool climate, take advantage of solar gain and give it a southern exposure. Be sure you build on a spot that has good drainage, for wet barns are bad for horses and tack . . . and people and dogs and cats. And while you're planning, consider that the road going to it will carry a lot of traffic and needs to be in good repair.

Because closed-in horses tend to get respiratory problems, one of the first things to consider when planning your barn is ventilation. Providing a window in each stall gives the horse a nice view while supplying light and air. Each window

◄As every horse owner knows, the road to heaven passes through a tidy barn with healthy, happy horses.

51

should be covered with bars to prevent breakage. Further ventilation can be provided by bars in the sliding stall door and across the front of the stalls, which also keep stabled horses from reaching out and nipping a horse being led in or out, or interfering with a horse standing in cross-ties in the aisle. Or you can use heavy mesh wire or chain link fencing instead of bars.

## Stalls

For ponies, stalls should be 10 × 10 and for horses 12 × 12. Broodmares and stallions should have larger stalls whenever possible. Stalls can be constructed of 2-inch oak or the like, or cinderblock or brick—any smooth, strong surface (with no protruding objects) that can withstand the kicking and rubbing that makes a horse a horse. The height of the stall should be at least 8 feet. The stall doors should be 4 feet wide and 8 feet high. Many people feel that some visibility between horses makes them happier and allows for better ventilation, so they make their stall partitions 5 feet high and put bars on the top. Other owners make one solid wall to the top of the stall. It just depends on you and your horse.

Although stall floors take a lot of wear and tear, they should not be made of hard materials, such as wood and concrete, that tire a horse's legs. Good materials for floors include well-tamped dirt, clay, or a 6-inch base of bluestone. These materials provide some drainage and are relatively inexpensive.

Aisles can be well-tamped dirt or earth with a sawdust covering, crushed bluestone, macadam, or broom-finished cement.

Ample lighting should be provided. A light in each stall, at various places over the aisle, and in the tack and feed areas should do the trick. Lights should be at least 8 feet high and incased in a wire sleeve or cage.

## Grooming Area

In small barns, cross-ties are often hung in the center aisle, which then doubles as a grooming area. It provides plenty of room to move freely on both sides of the horse with easy access to the tack room for whatever is needed. Some barns have either a broom-finished (non-slip) center aisle with a drain or a separate, 12 × 12 wash area where horses can be bathed. Horses can certainly be washed outside, but with a revolving horse in one hand and a bucket or hose in the other, you do tend to make mud puddles, tear up the grass, and go back to the barn with six very muddy feet. Wash areas are a great addition to any barn.

## Hay Storage

Once you've determined how many horses you intend to keep, you can compute how much space you need for storage. Figure that a bale of hay requires about 10 cubic feet. Decide how often you're willing to restock your hay and how much you'll feed. For instance, if you have two horses eating a bale a day between them, 900 square feet of hay storage space would store 90 bales, enough for three months. If you bed with straw, you'll have to store that, too.

If you decide to build a hayloft, your builder will construct it on the basis of the maximum number of bales you need to store. He's figuring for both bale weight and air circulation. To store more than your original intention could stress the floor or start a fire. Lofts should always have plenty

Photo by Margaret Thomas

◄Mucking the stalls may not be the most enjoyable part of owning a horse, but it's a chore that has to be done.

removed at certain intervals during the year. If you advertise in the paper, you may find gardeners who will haul it away for you. If you have a manure spreader, the problem is solved. Just be sure you don't spread fresh manure on horse pastures, for it contains the eggs of parasites which will reinfect your horse. (Cattle and sheep are not subject to equine parasites.)

## What You'll Need

For easy daily maintenance, a well-managed stable needs the following items to keep the horses healthy and clean and the barn tidy and well organized.

- A leather halter will require the same kind of care that all leather does, but it's preferable to nylon. Nylon halters get stiff and rub the horse, and they don't break. If your horse catches his halter on something, he'll pull back hard. It's better to have a broken halter than a broken neck.
- Three cotton shanks (lead ropes)—two for cross-ties and one to lead the horse.
- A well-fitted, all-weather blanket for clipped horses, sick horses, and cold climates.
- A box containing grooming tools such as brushes, sponges, a sweat scraper, and a hoof pick.
- Other necessary tools include a 3-tine hay pitchfork and a 5- or 6-tine manure pitchfork, a wheelbarrow, a broad, flat shovel, a broom, a rake, a hose, and a frost-free hydrant.
- Each stall should have a removable water bucket and a feed tub. A hay rack is optional.
- The tack room should contain three or four sponges, a choice of saddle soaps, a

of ventilation. Louvers in each end of the loft and a wind-operated turbine draw air out and require no electricity. Haylofts should never have skylights, and the roof should be a light color.

## The Manure Pile

The manure pile is usually stashed behind the barn, far enough away for fly control but close enough for easy access. Some towns have ordinances requiring that manure be put in a deep, covered pit and

►A working fire extinguisher should be located in every barn where it's easy to lay your hands on it if necessary.

tack hook for cleaning bridles and girths, and a rack for cleaning saddles if you wish.

- Medical supplies are usually stored in the tack room. They should be kept in a covered box or cabinet away from rodents, dogs, and children.
- All barns should have no-smoking signs on the doors and an accessible, working fire extinguisher.

## Feed Bin

Besides storing grain, the most important jobs of the feed bin are to prevent contamination of the feed from rodents, and to withstand the strength and perseverance of the cleverest horse trying to open it. A wooden bin will soon become a rodent motel if it's not lined with tin. Some people use metal garbage cans to store feed. They don't have high-volume storage capacity, but they're cheap and mouse-proof. If you use them, be sure they're out of reach of the occasional loose horse. (A loose horse, spending the night eating grain, will most likely founder.)

If you can get your hands on an old milk cooler, it will make a perfect feed bin. Mine stores about 800 pounds of feed, the metal lid is much too heavy for a horse to open, and the frustrated rats don't have a chance.

## Bedding

Bedding provides cushioning for a horse's feet and absorbency for urine and droppings, as well as warmth in the winter. There are several kinds of bedding available, depending on the part of the country in which you live.

*Straw* is very absorbent—the Cadillac of stall beddings—but it's expensive. You'll need about one bale per stall per day for stabled horses. Smaller barns may have a problem storing it, for it must be kept under shelter and not allowed to mold.

*Sawdust* is a very good absorbent bedding which can be picked up or delivered by your local sawmill. It's cheaper than straw and may be stored outside under a tarp. It's convenient, economical, and relatively dust-free. There is, however, one serious problem to avoid if you use sawdust.

There is an ingredient in black walnut trees that is highly toxic to horses. A few years ago a local sawmill bought a stand of walnut trees (which are native in the mideastern and midwestern states). Many area horse owners rely on this mill for bedding, and one day several truckloads of black walnut sawdust were delivered to various barns. Within just a few hours some owners had called their vet to report mysterious lamenesses—but by the next morning everyone who had bedded with it had called. About 90 percent of the horses who had stood on the bedding were lame, and some were in severe pain with acute laminitis. It is not known what the toxic element is, but it's trouble. Whenever you call for a sawdust delivery, make sure you're not getting black walnut.

*Wood shavings* aren't quite as absorbent as sawdust, but they get the job done. Be sure to remove any larger pieces of wood that have sharp edges. You can order shavings from your feed store, which usually delivers them in large, heavy paper bags.

*Sand* is often used as bedding. It's cheap and absorbent, but it doesn't provide any warmth in a cold, damp climate. If you use it, be sure your horses don't eat it (feed them their hay in a rack), for it can cause intestinal blockage and colic. Never bed with beach sand because horses like to lick it for the salt content.

Although straw, sawdust, and shavings are the beddings most people use, you can also use peanut hulls, peat moss, or pine needles.

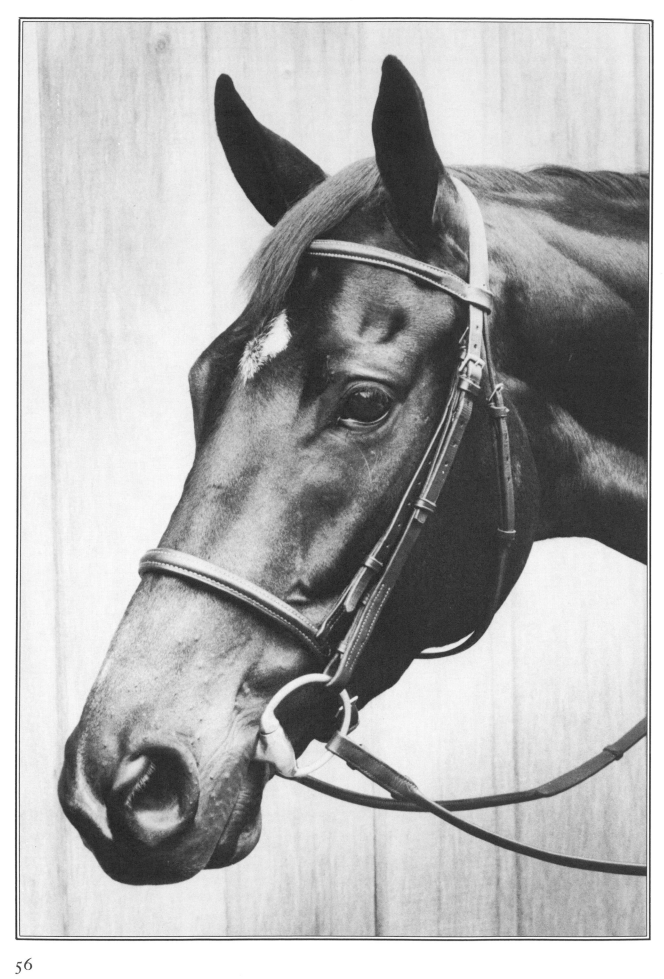

# The Tack Room

—◂•▸—

The word "tack" refers to a wide range of gear that horse people use to ride and train the horse. The tack room can be as simple or as fancy as you want to make it. What you need is a place to hang your tack (the collection continues to grow over the years) in a way that it remains clean, well supported, and safe from tack gobblers such as rats, mice, and the family dog. A lock on the door will give you some security when you're away.

The ancient tack room contained no more than a single leather thong to tie around the horse's lower jaw; the excess length of leather was used as a rein. That was the way the American Indians rode. The Egyptians wanted more control, so they designed the snaffle bridles whose likenesses can be seen carved on tombs dated around 1600 B.C. and which bear a remarkable resemblance to bridles hanging in today's tack rooms.

But that is not the end of it. Today's tack is often quite sophisticated, and it is designed to aid the rider in getting still more perfect performance from the horse. Some of it may be a little trendy. Some of it may seem a bit mysterious. There is a lot to choose from, and all of it is expensive. But don't let a trip to the tack store overwhelm you, because the family rider can limit himself to the basic necessities for everyday use and be perfectly well equipped. You will need a bridle, a saddle, a saddle pad, a girth or cinch, stirrups or irons, and stirrup leathers.

In this chapter we will discuss what to look for, how it should fit, and how to keep it in good condition.

## THE BRIDLE

The bridle is composed of the headstall, the browband, the noseband or cavesson, the cheek piece, the rein, and the bit. Some Western bridles omit the noseband or browband. Most bridles are fairly similar in function, but they vary in design, quality of leather, and stitching. The functions of the bridle are to control the horse's head, to encourage proper head carriage for the type of work the horse is doing, and to aid in changing the direction and the speed of the

◂ Like this bridle, every piece of tack has a specific purpose and must be fitted and adjusted properly to the horse.

horse. The influence the bridle has on the horse is directly related to the choice of bit that we put in his mouth.

There are four categories of bridles or bits. They are the snaffle, the curb, the pelham, and the hackamore.

Each type of bit has several variations that are designed to influence its severity and how it works. The wider, softer, and smoother mouthpieces give a wider, milder distribution of surface pressure in the horse's mouth. And, of course, the converse is also true.

The bit lies on the "bars" of a horse's mouth, a sensitive gum area between the molars and the incisors given easily to bruising and permanent nerve damage.

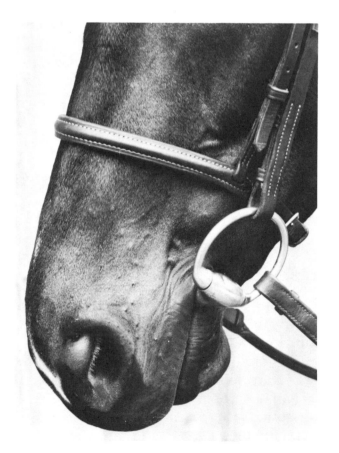

To be most effective and least objectionable to the horse, the bit must fit correctly in the horse's mouth.

Depending on the choice of bit and how it is made, the other pressure points on the head that a bridle affects are the poll, the nose, the chin groove, the lips and the tongue, and the roof of the mouth.

A well-fitted bridle looks neat and tidy on a horse. The browband shouldn't drop down over his forehead or pinch his ears. The noseband should fall below the cheekbones with about 1 inch of space between the bridle and the head. The throat latch should be pretty roomy, with space enough to allow for breathing and flexing of the head, and the cheek piece shouldn't flap when leverage is applied to the reins.

A novice rider, whose balance and strength are not yet developed, often uses the reins as a means of support. Though it is a natural step in learning to ride, this can hurt the horse, who will show his discomfort in a variety of ways, from tossing his head to rearing. For this reason it makes sense to begin and stick with the mildest form of bit you can use to manage your horse.

## The Snaffle

In its simplest form, the snaffle is the mildest of all bits and is always used with one rein. It puts pressure on the bars, the tongue, and the lips of the horse's mouth. It has various designs that alter its pressure points.

## The Pelham

The pelham bit requires two reins, and it combines the curb and snaffle into one bit. The upper and wider rein controls the milder snaffle, and the lower, thinner rein is to control the curb. This bit is always used with a curb chain. One of the mildest

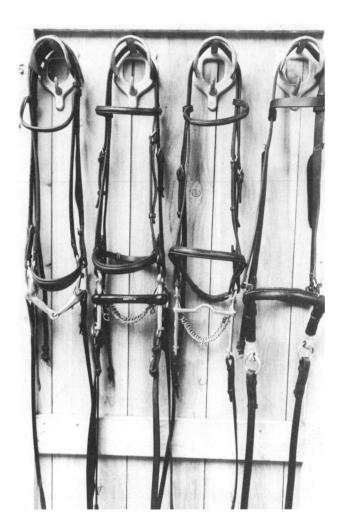

pelhams is the tom thumb, which has a wide rubber mouthpiece and a very short curb shank.

## The Curb

The curb bit also uses one rein, but it is made to use with a curb chain or a curb strap, which is fitted against the chin groove. As is true of the snaffle, the curb has many variations that alter its severity. The curb operates on the poll, the chin groove, and the bars.

## The Hackamore

The hackamore, or bosal, is a bridle altogether different in concept, because it has no bit. It operates by putting pressure against the horse's nose. It is used in both English and Western riding and is often seen on jumpers. It rests about 4 inches over the horse's nostrils and must be used with respect, for, as any horse will tell you, he has a very sensitive nose. This bridle is a good choice for a horse with a damaged or fussy mouth.

# THE SADDLE

There are many kinds of saddles, but generally they fall into one of two categories: English (flat) or Western (stock). The question of which one you want depends on what kind of riding you want to do and what type of horse you buy, rather than where you live. There are plenty of people who ride Western in the East and just as many who ride a flat saddle in the West. Timid beginners may like the security of a deep-seated stock saddle, which has a horn

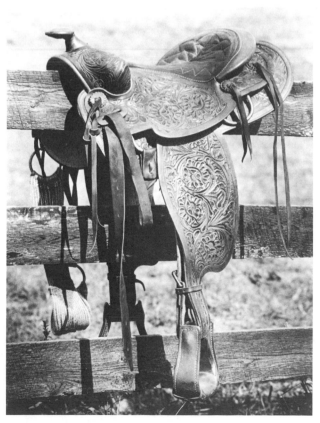

►The horn on a Western saddle is not meant for beginning riders to hold on to, but for working cowboys to hang their ropes on.

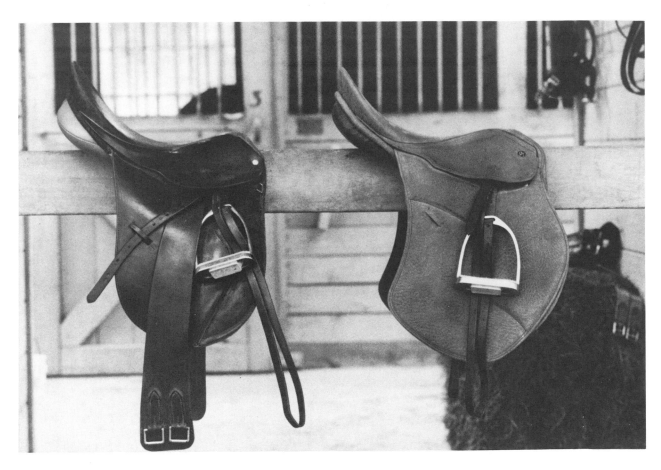

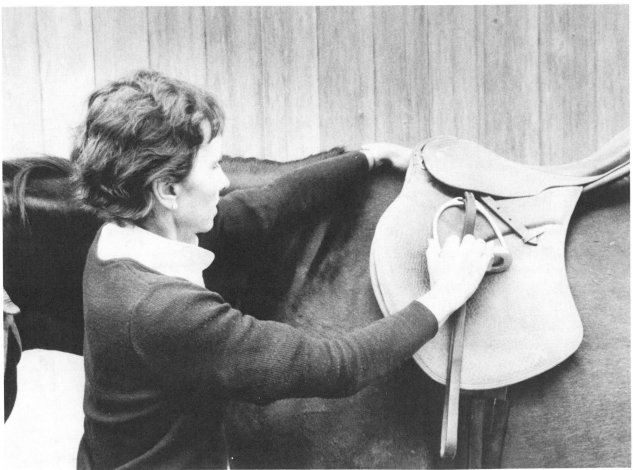

to hold on to, but in my opinion, if you learn to ride in a flat saddle, you can transfer to a stock saddle more easily than the other way around. If you intend to learn to jump, then you will ride in a flat saddle.

When picturing a Western saddle, many people see the image of the horn with a coiled rope dangling and silver-laden, tooled leather cradling the head of a worn-out cowpoke sleeping in the moonlight. Western saddles connote romance and adventure, a piece of the past slipped by, rejuvenated occasionally by Clint Eastwood galloping across the screen hell bent for leather. When thinking of the English saddle, the simple picture of a small, un-adorned saddle without a horn comes to mind. But there are various types of English saddles. To describe a few, the dressage saddle has a long flap to accommodate riding in a long stirrup, jumping saddles are designed for a sharper knee angle, and racing saddles have a tiny surface area (since jockeys spend most of their time crouched over their horse) and weigh practically nothing. For most beginners, a good all-purpose saddle is most likely to give the best service.

## New or Used

Well-made new saddles are very expensive, but worth the price if they are within your budget. However, if your choice is between an inexpensive new saddle, say an Argentine saddle, and a well-made old one in good condition, the second-hand saddle will probably be the best buy in terms of fit, durability, leather quality, and workmanship. It is as true of saddles as of many other things: they just don't make them like they used to.

When you find a saddle you like, ask the seller to allow you to try it on your horse. Since one horse's back is different from another's, and saddles are all cut differently as well, it is important to be sure that the saddle fits properly. There should be about 2 inches of space between the saddle and the horse's spine. The weight of the saddle should be distributed evenly over the rest of his back, no area receiving pressure that can cause muscle fatigue or chafing.

If the saddle fits the horse properly, do up the girth and get on. Sit on your seat bones in the deepest part of the saddle with your legs hanging down, muscles relaxed. Now, adjust the stirrups so that the bottom of the stirrup is level with your ankle bone. With your feet in the stirrups, heels down, there should be plenty of room for your knees, and you should be able to fit your hand between your seat and the back of the saddle. You must feel comfortable, as though you are part of the horse, rather than perched over him. It's a good idea to have an experienced rider check you out. If everything seems to jell, the fit and the feel are good, then you are in business.

With proper treatment and perhaps a little new stitching over the years, your saddle should last a lifetime. It should be kept on a saddle rack that gives some support to its tree, or foundation. You can buy a rack or make your own. One useful idea for a saddle rack is to fasten a large mailbox to the wall: the saddle sits on top, and the inside holds grooming and tack-cleaning equipment. When several people are using a barn, it's a good way to keep your gear in one private place. Bridles should be hung by the headstall (the upper part of the bridle that goes behind the

horse's ears) on a rounded support, not a nail. Tack-shop bridle racks are great, but a coffee can nailed to the wall will do.

## The Saddle Pad

The saddle pad has two purposes. It protects the horse's back from friction and pressure, and it absorbs sweat. There is a wide range of pads, and you should be able to choose one that is suitable for both your horse's back and your pocketbook. They range in price from about $5.00 to $80.00. To choose the correct one, take a look at the horse's conformation and consider what kind of work he is going to do.

Some horses have sensitive backs to begin with, so they need more protection. Narrow horses with a prominent withers and thin-skinned horses such as Thoroughbreds and Arabians are more likely to develop sores and sensitive areas than mutton-withered, wide horses whose backs distribute saddle weight more evenly. Horses carrying heavy riders should have extra protection, and, of course, horses undergoing long hours of work, such as endurance riding, where friction and sweat are constant irritants, need extra padding.

Whatever you choose, be practical, for pads should be washed frequently. Machine-washable pads make a lot of sense, and manufacturers are marketing really good synthetic fabrics these days. There are wipe-clean, foam rubber pads (cheap and absorbent, but not very long lasting), single- or double-ply quilted cotton pads (double is better), fleece pads, and a breathable vinyl model that is excellent. For Western riding, a washable Navaho-style blanket is practical and popular under a stock saddle, and it is seen more and more under flat saddles as well.

The heavier the fabric, the less likely it is to slip around under the saddle. A pad that pleats and bunches up may cause more problems than no pad at all. Be sure to get one that will stay put.

## Girths

There are several kinds of girths to choose from. Some are made of leather, some from string or hemp. Leather is used most often, but string girths are nice to have on hand, for the soft nylon is easy on an unfit horse whose skin is soft and subject to chafing. Be sure that the girth you buy is of good quality with unworn stitching and secure buckles, for girths take a lot of stress.

## Stirrup Leathers

Stirrup leathers are made of various kinds of leather, including tough buffalo hide. Be sure the stitching is intact and carefully done. New ones tend to stretch a bit, so you may need to make a few adjustments in the length of your stirrups from time to time.

## Stirrup Irons

Stirrup irons come in two materials, nickel and stainless steel. Nickel is, of course, a little cheaper, but it will rust and it may break. Stainless steel will last forever and is safer in the long run. When you buy your irons, wear the boots you intend to ride in and make sure you allow about 2 inches of space between your foot and the inside of the stirrup for easy disengagement.

▶ Particularly good for young children and beginning riders, the safety stirrup will disengage if the rider falls.

No one should ride in stirrups that fit the foot snugly. It's an accident waiting to happen.

## The Safety Stirrup

The safety stirrup is made with a rubber release on the outside. It is a simple device that ensures disengagement of the foot from the stirrup in case of a fall. Safety stirrups are good for all novice riders and particularly for youngsters.

## Martingales

Some horses who carry their heads too high may need a martingale (or a tie-down) to ensure proper head carriage. There is a choice between a standing or a running martingale. A standing martingale has a yoke or a strap which is fitted around the horse's neck just in front of his withers; a

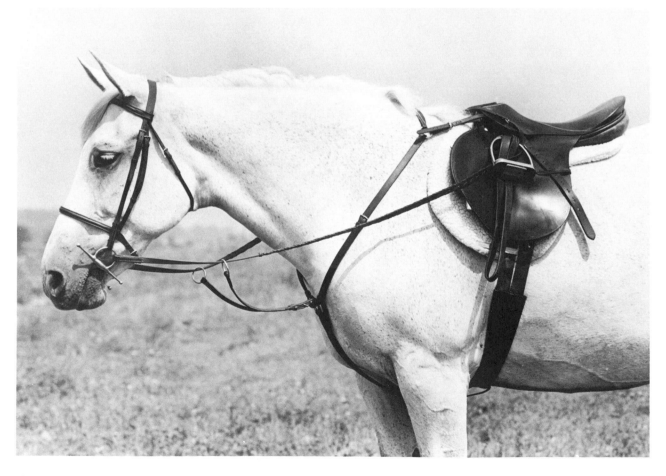

second strap attached to the underside of the noseband of the bridle passes through the yoke and down between the horse's front legs to the girth. When the horse is standing naturally, it should be long enough to reach his throat, thus providing enough length for normal flexion of the head, but short enough to prevent the horse from raising his head high enough to bash into yours, an unpleasant meeting of the minds, to say the least.

The running martingale also has a yoke around the neck with a strap through which the girth passes, but it differs from the standing martingale in that the leverage used to keep the head down is applied not to the noseband but to the reins. The rings through which the reins pass should be long

enough so that they can reach the horse's withers. This adjustment allows for ample free movement of the horse.

## The Crupper Strap

The crupper strap is a handy piece of tack which is as amusing as it is practical. Some ponies take advantage of their fat undefined withers by occasionally ducking their head and sliding the saddle, kid and all, down their neck. This specialty is usually performed going downhill, the faster the better, and ends in a sliding stop which sends the rider flying over the pony's head. The crupper strap simply attaches to the back of the saddle, slips under the pony's tail, where it is padded, and loops back to the saddle to hold it in place. End of trick.

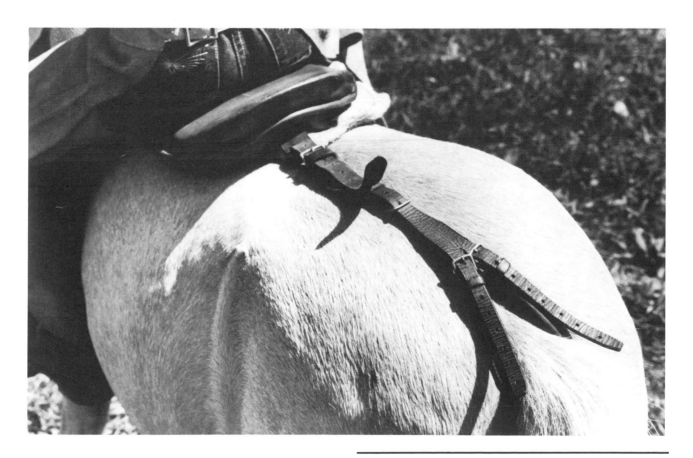

A simple but effective device, the crupper strap keeps the saddle from slipping forward and the child from sliding down the pony's neck.

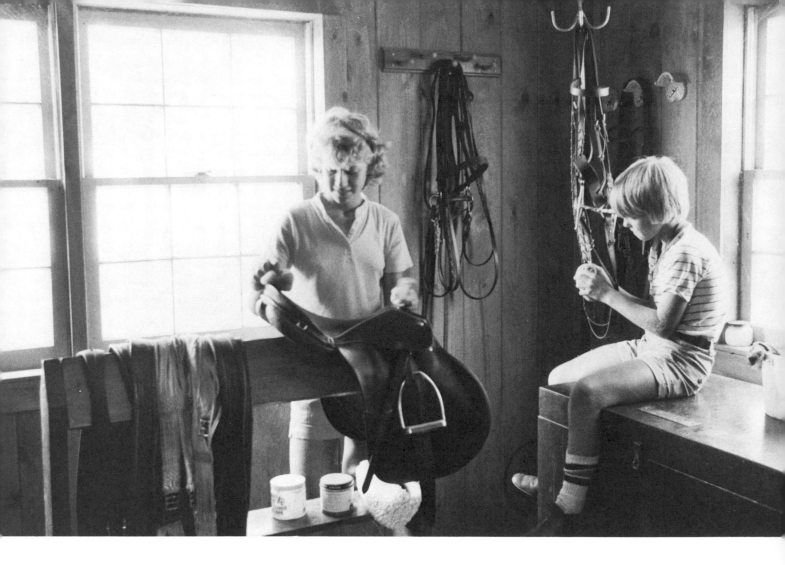

## To Clean Tack

Cleaning tack is a twofold process. The first step is to remove the sweat and dirt from the leather. You will need a bucket of water and a sponge for this. Some people like to pour a little ammonia in the water to help cut the dirt and gummy sweat. Unbuckle all parts of the bridle and remove the reins from the bit, and wipe everything clean with a damp sponge. Keep rinsing the sponge and change the water when it gets dirty. When the dirt is removed, proceed to the second step. With a clean, dry sponge, use a good saddle soap or leather conditioner to restore the oils, suppleness, and shine to the leather. There are many products to choose from, but Lexol, glycerine, and Murphy's Oil Soap are the real staples, and you can't go wrong using them. I like to dampen the sponge with Lexol and then rub it with a bar of glycerine to get a dry lather. An application of this brings the leather back to life and gives it a smooth, dark shine.

Be sure to clean and condition all unused tack once a week. It gets dirty just hanging in the barn, and the leather becomes dry and brittle.

After a morning's ride in the rain or a trot across a stream, your tack may be soaked. Let it dry naturally, away from sun and heat. It may not dry as fast as you would like, but the leather will be a lot healthier if you do not try to hurry the process. When it is dry, clean it and oil it lightly. When it is supple again, just carry on with your normal routine.

## To Recondition Old Tack

Occasionally one finds an old bridle at a sale or in someone's attic that has dried out. The standard procedure is to use neat's-foot oil to recondition it, but some people like mink oil, baby oil, or olive oil. Whatever you use, warm the oil, or at least have it at room temperature, and let the bridle soak in it for an hour or so. The problem is to get oil back into the leather without saturating it to the point that it will ooze oil all over you and your horse for the next week or so. Also,

too much oil tends to rot the stitching. After the leather has absorbed the oil, wipe off any excess with a clean sponge and let it dry. Next day, repeat the process if needed, or finish the reconditioning with a good saddle soap.

## To Stain New Tack

If you buy a new piece of tack which you care to darken, you can use several light applications of neat's-foot oil, or, to make your work shorter, you can mix it with a commercial leather stain. Just rub it on with a clean sponge once a day until the color suits you. Some tack stores provide a staining service which is usually inexpensive and sometimes free. 🐎

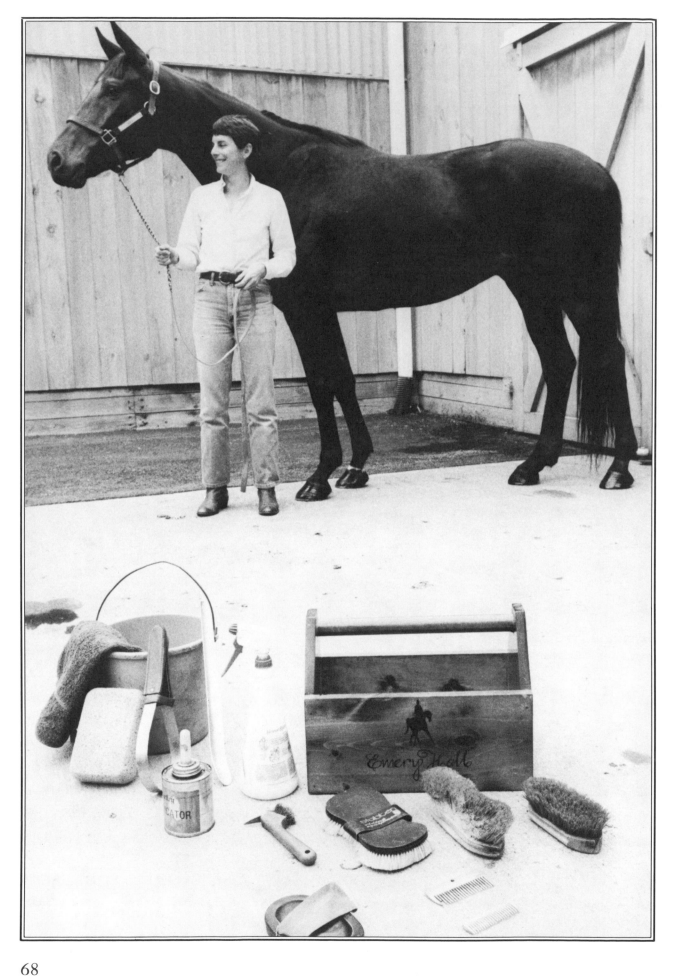

# Stable Management

Stable management includes keeping the barn in a tidy state and following a routine that monitors and manages the horse's health, feeding, and exercise. Good stable management also requires awareness of the horse's state of mind.

## A Proper Balance

A knowledgeable, sympathetic trainer uses the horse's mind to his advantage. He encourages the horse to enjoy his training, which, to a large measure, has to do with allowing the horse free time to express himself as a horse—to graze, to be part of a herd (or at least to have a companion, perhaps a cow or a goat), and to exercise his instincts as well as his body. Deny him that, and he will become depressed or nervous. Most vices and bad habits center around an imbalance of regulation and freedom. A good manager will observe the horse and respect him, tailoring training and turn-out time to the needs of each horse.

In general, during the summer, horses are kept in the barn in the daytime, to protect them from heat and flies, and are turned out at night. In winter, they are let out to catch the warmth of daytime sun and protected at night. However, very high-strung horses may need more turn-out time and more exercise. Some horses do just fine with an hour a day in the paddock. Observe your horse closely, and he will give you all the clues you need to make the proper decision.

## GROOMING

Stabled horses should be groomed daily, before and after riding. Grooming keeps a horse healthy by stimulating circulation and improving muscle tone. Brushing distributes skin oils throughout his coat. A horse breathes partially through his skin, much like you and I do, and a clean skin, free of dust, parasites, and disease, facilitates this process.

You don't have to do the entire job before you go riding, but you should take care of a few things:

- Use a hoof pick to clean his feet, and then apply a lanolin hoof dressing.
- Be sure that his back is completely clean and dry, to avoid saddle sores. Saddle pads should also be clean and dry.

◄ Brushes, sponges, and other cleaning equipment can be kept together in a box. Their frequent use will produce a clean, happy horse .

69

- With a *soft* brush and rag, clean his face and head. It'll help keep your bridle clean.
- Brush his legs with a body brush and be sure to clean the back of his pasterns.

This process can be done in a few minutes, and, therefore, it's labeled "knocking him off." Go on your ride and enjoy yourself, because when you return you've got some real work to do.

## The Entire Job

To thoroughly groom your horse you'll need the following tools:

dandy brush (stiff)
body brush (medium)
water brush (soft)
rubber curry comb
bucket
lanolin-base shampoo
2 sponges
scraper
terry towel
hoof pick
hoof dressing and brush
comb
scissors
clippers (optional)
fly wipe or fly repellent

Begin by putting the horse in a halter (never a bridle) and standing him in a well-lit area in cross-ties. In grooming, as in all other cases of handling a horse, always begin on the left, or the "near," side.

**STEP 1.** Starting at the top of the horse's neck, use the curry comb in a circular motion to loosen dirt, dust, and manure, and brush it off with the dandy brush.

►Standing the horse in cross-ties in a well-lit area makes grooming him much more manageable.

Continue down his neck and over his body. The saddle probably caused him to perspire, and his withers and girth area must be perfectly free of sweat. If you need to, use a little warm water and a sponge over this area. Continue with the curry comb and dandy brush, removing dirt and mud. Do not clean tender spots with either of these tools—his head and legs are too sensitive. There's an old prayer said to have been spoken by a horse long ago:

Curry softly,
Curry please,
But never curry
Below the knees.

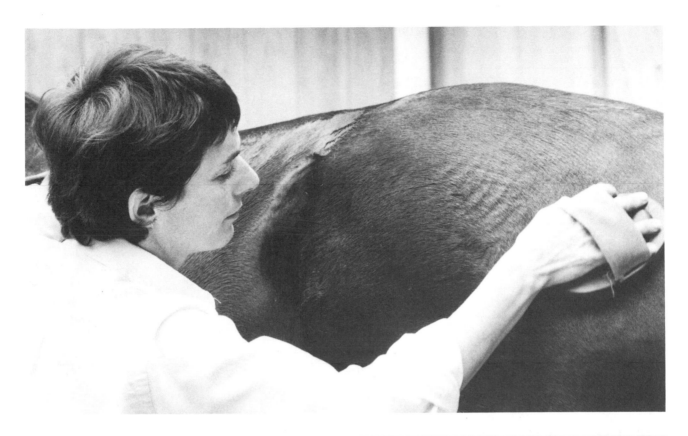

**STEP 2.** The next step is to pick up the body brush with your right hand and begin the process again, this time using the curry comb to clean the brush. Every few strokes, run the brush across the curry comb to clean it, and then clean the curry comb by tapping it on the wall. (Yes, it's tiring—just think of it as good exercise.) The body brush is soft enough for the horse's legs. Now, your horse is about half clean.

**STEP 3.** Remove any remaining grass or manure stains with a sponge and warm water containing a lanolin-base shampoo. This is a good time to clean the underside of his tail (the dock) and his socks, if he has any. And while you are half kneeling, wipe a mare's teats and a gelding's sheath *very gently*. Then wash any mud off his hooves. Rinse carefully. With a separate clean sponge reserved for this purpose, carefully wipe his eyes, ears, and nostrils.

A "fly scrim" is a simple device that can help protect your horse from the constant annoyance of flies in his face.

**STEP 4.** The next round requires the water brush and terry cloth. Brush with the right hand and follow the brush strokes with a swipe of the cloth. This will smooth his coat and remove surface dust while distributing oils through his coat. Use this soft brush on the horse's head, too. Now, untangle or pick the tail with your fingers, and then comb the mane and tail, following up with the slightly dampened water brush.

**BOT EGGS.** If the horse has any bot eggs (see page 77) attached to his legs or body, remove them by wetting them with warm water to loosen them, and scrape them off with a sharp knife or rubbing alcohol.

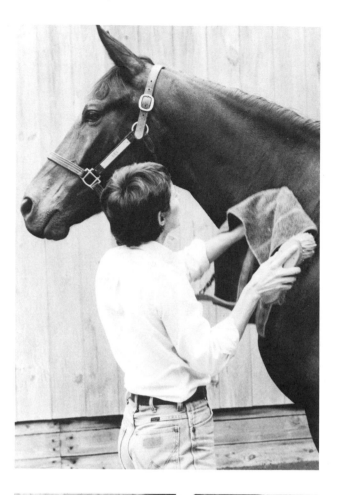

**FEET.** Using a hoof pick, clean the horse's feet. You'll be removing bedding and dirt and perhaps even small stones that pack up into the foot and cause pressure (which slows down foot circulation), and you'll be removing bacteria as well. Now apply the hoof dressing liberally, right up to the coronet band.

**WHISKERS AND FETLOCKS.** From time to time you will want to trim your horse's whiskers and fetlocks with scissors or clippers. The breed of your horse will dictate how long his mane and tail should be. Find out what tradition dictates and follow the custom.

**FLY REPELLENT.** In fly season, horses are pestered continually by armies of flies. To protect your horse, finish the grooming by rubbing or spraying some sort of commercial fly repellent on him. Follow the directions exactly. Do not think that more is better. Remember, you are using a chemical. I call the more-is-better approach killing

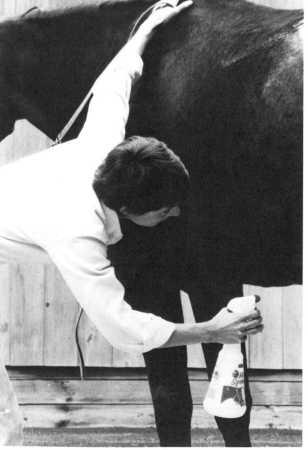

►A little fly spray after grooming is an easy way to help keep your horse protected from a terrific pest.

your horse with kindness. A student of mine came quite close. He once saturated his horse with undiluted fly repellent which the horse absorbed through his skin. The horse nearly died of nerve damage. It is important to follow directions *exactly* when using any chemical or medicine on your horse.

## Washing

After a summer ride, your horse may come back sweating and lathered up, which calls for a bath. Use three buckets of warm water (you can simply place them in the sun before you ride to warm them). Squeeze a little lanolin-base shampoo into one bucket, and, using a large sponge, wash the horse all over. Rinse him with the two buckets of clean water. Remove excess water with the scraper, and towel him as dry as you can; then let the sun do the rest. When he's dry, brush with a body brush and a water brush, following each stroke with a towel.

# THE HORSE'S FOOT

The hoof is a remarkable creation. It is rigid enough to withstand thousands of pounds of shock and flexible enough to absorb and divert the shock. At the same time, it stimulates the blood supply and protects the ligaments and bones within it. According to hoof researcher Dr. Douglas Leach, the average 1,000-pound horse, while standing, is carrying about 300 pounds on each front leg and 200 pounds on each hind leg. And that's the least of it. When galloping a four-beat gait (meaning that each foot takes a turn at supporting the horse's entire weight during a stride), speed increases the pressure to as much as 2,000 pounds per

hoof. When landing from a jump, the front feet are supporting even more weight. The equine hoof really takes a beating, and we need to know how to take care of it.

## Does Your Horse Need Shoes?

What you ride, how you ride, and where you ride will determine whether your horse needs shoes to begin with. Many ponies, for instance, have very dense hooves and may not need shoes at all, or only in front. Arabians and Appaloosas have particularly good feet, too. On the other hand, gravel roads can break up the hardest hoof. But even if your horse is working on soft terrain and has a good foot that resists breaking, he will need to see the farrier about every six weeks for trimming, because a horse's foot grows continuously. The normal hoof re-

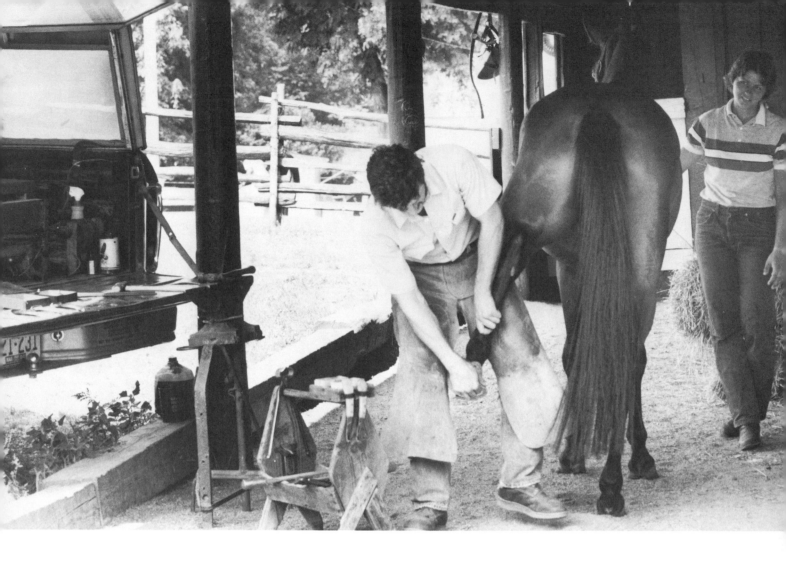

places itself, from coronet band to the ground, in about a year. Because of that continuous growth, shaping and balancing are necessary to avoid splitting. Just like your fingernails. And because the foot is of such primary importance, it must be cared for by a first-rate farrier.

## Getting a Farrier

A farrier is a professional who trims and shoes horses' feet. Most good farriers spend a lot of their time shoeing horses at a professional barn—show stables, race tracks, polo clubs, and the like. But there's no rule stating that they won't work for small family operations as well. Inquire around and sooner or later you'll find one. Since good

farriers have skill akin to an artist's and take great pride in their work, they're worth their weight in gold. Be nice to your farrier, for in the emergency of a pulled shoe or torn wall, he holds the key to whether or not your horse leaves the barn.

A good farrier can improve or correct various problems pertaining to how your horse moves. For instance, forward movement in a horse who toes in or out may cause strain on his legs or "brushing" (knocking one front foot against the other). By properly balancing the foot, a farrier can improve or eliminate this condition. At the trot, some horses continually hit their ankles with a hind leg, known as "forging." By shortening the toe and altering the angle

▶ Using a longe line to exercise your horse is an alternative to riding him or giving him pasture time.

of the shoe, a good farrier can probably relieve this problem. Some young horses who are on their way to future leg problems because of slightly unbalanced hooves, can be put straight in the hands of a good farrier. He can relieve pressure on sore spots and add support to areas that need it. Half artist, half mechanic, a good farrier will help keep your horses sound.

There are three basic steps to shoeing. Each step requires careful measuring, a steady hand, and a good eye. First the feet are trimmed and leveled. Then a shoe is made to fit each foot exactly (rather than changing the foot to fit the shoe). And, finally, the shoe is nailed on.

After a little study on what a properly shod foot should look like, you can get an idea of how well your horse is shod by checking to see that the foot rests evenly on the shoe; the nails are spaced evenly, in a straight horizontal line, and well clinched; the axis of the hoof merges with the pastern to form one straight line; and the horse trots out soundly, moving at least as well as he did before shoeing. If these things check out, your horse may well be in the hands of a good blacksmith.

Proper shoeing is not all that's required for good foot care. There are a few things you should do for your horse at home:

- Pick out your horse's feet at least once a day.
- Brush on a hoof dressing that contains lanolin.
- Keep an unshod horse off gravel roads or other areas with small, sharp stones. They will bruise his feet.
- Be sure your horse gets plenty of exercise to stimulate his circulation.
- Maintain a good feeding program.

## TEETH

Millions of years ago, when the Ice Age had altered the terrain, and tough grass replaced tender leaves as the horse's diet, nature changed the horse's jawbone. It became thicker and therefore better able to hold stronger teeth that would grow continuously to compensate for grinding away during mastication. Since the grinding down and the growth generally take place at the same rate, everything should be okay, right? Wrong.

Horses grind their food by sliding their molars from side to side, producing razor-sharp edges on the insides of the lower teeth and the outsides of the upper teeth. Edges on the inside can bother or cut a horse's tongue, and outside edges cut the cheek. The result is that eating becomes a painful chore, and the horse loses condition.

### How To Know

There are several ways to find out if your horse needs to see an equine dentist or a vet. One way is to feel his teeth. Facing him, feel the upper molars on *your* left side with the thumb of your right hand. Run your thumb *carefully* over the outside edges of his teeth. The hazard to avoid here is getting bitten or getting a sliced thumb from the edges of his teeth.

You can also get some idea of the condition of his teeth without risking your thumb. If he spills a lot of feed as he eats, or eats with his head turned to one side, or if you see whole grain in his droppings, you may conclude that your horse needs to have his teeth checked.

"Floating" is the process of removing the sharp edges of the teeth with a special rasp. It is done by an equine dentist or by a vet. Most horses need to have their teeth floated once a year.

## PARASITES

Horses are subject to many kinds of internal parasites which can infest and damage most of their organs. Since the parasites travel throughout the horse's system, and worming only catches parasites in the stomach or intestine, most horses have worms all the time. However, if we do not regularly reduce the population of parasites, they will cause dysfunction or death.

Repeated burrowing of parasites through a horse's organs causes internal bleeding, which leads to ulcers and anemia. This burrowing also produces scar tissue throughout the horse's system, narrowing the arteries and decreasing the supply of blood to his intestines and legs. Loss of appetite, a pot belly, lethargy, poor condition and loss of shine to coat, a low-grade fever, or repeated colic can indicate that your horse is ridden with parasites.

*Large strongyles* (blood worms) are the most insidious parasites of all. Begin by understanding that the word "large" refers to their sucking capacity. There is also a small strongyle. Strongyles burrow constantly through the lining of blood vessels and bowels, causing scar tissue, ulcers, bleeding, anemia, and partial or total blockage of blood supply to the legs or intestine, which often leads to death.

*Pinworms* live in the large bowel of the horse. The female migrates to the anus to

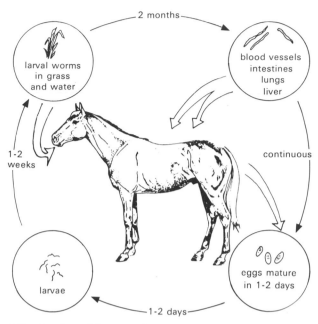

**Life Cycle of Strongyles**

lay her eggs, which cause irritation and itching. The horse backs into the stall and frantically scratches his tail, sometimes rubbing off all the hair.

*Large round worms* typically affect yearlings. The foal swallows eggs which hatch to larvae and penetrate the bowel wall, proceeding to the liver and lungs by way of the blood stream. They return to the bowel to mature and reproduce, at which point the eggs return to the ground in the manure to start the process again. Round worms are cream-colored, thick, and up to one foot in length. Because of their size, the greatest danger they create is death from bowel rupture.

*Large and small stomach worms* enter the horse orally or by infesting a sore, which becomes a "summer sore" and is very hard to heal until frosty weather kills the larvae. Those that get into the stomach lay eggs which pass through the horse in his droppings, at which time the eggs are swallowed by the maggots of flies. As the maggot develops into a fly, he dines on the horse's skin, dropping larvae where either the horse ingests it or it gets into a cut—thus starting the process again.

*Tapeworms* live in the small intestine and damage the lining by repeated burrowing. *Bot flies* lay small (pinhead-size), orange eggs on the horse, commonly on his legs and belly, toward the end of summer. When the horse touches the eggs with his mouth, larvae enter and migrate to the small intestine, where they may live for about a year, producing deep pits at their points of attachment. They may cause pain (colic) or death. Bot eggs should be scraped off the horse with a sharp knife or rubbing alcohol, and worming should take place after the first few hard frosts.

Photo by Margaret Thomas

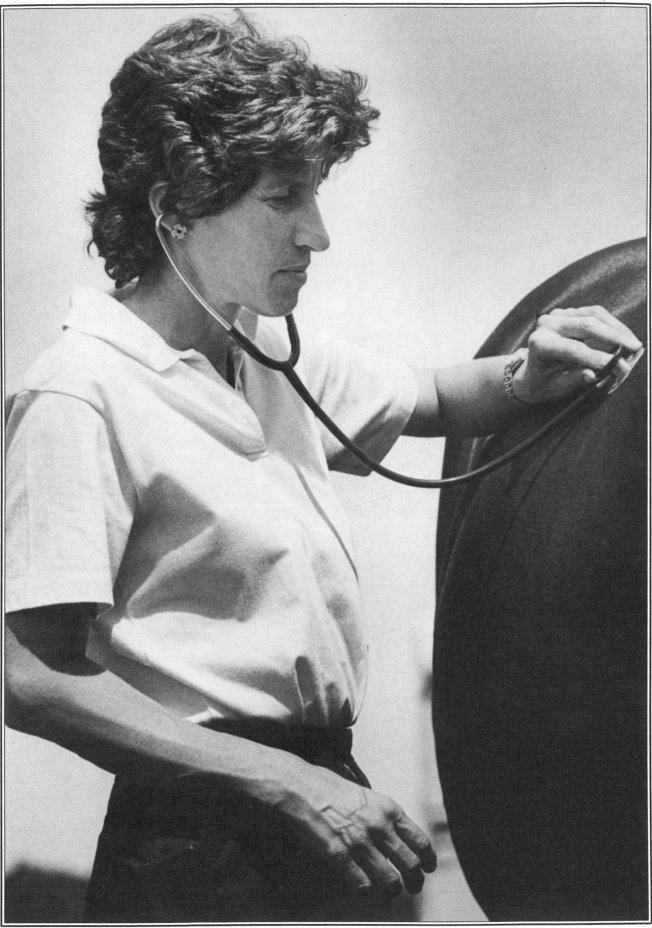

# Basic Medical Problems

Even in the best of situations, horses (like children) manage to bump and bruise themselves while playing, or suddenly come down with colds and flu, lose their appetites, run a fever, and feel depressed or listless. In this chapter we'll look at some of the health problems that are most likely to occur with the family horse, and talk about how you can recognize them, what to do about them, and when to call the vet.

Depending on where you live (what sort of climate you have) and what illnesses are concentrated in your area, you may want to discuss with your veterinarian the following inoculations: tetanus, encephalitis ("sleeping sickness"), strangles, rabies, influenza, virus abortion for mares, and botulism for foals. When your horse is inoculated, remember to make a note of the date on your calendar for future reference. It's a simple thing to do, and you'll thank yourself for being so efficient later on.

Since horses can't tell us where it hurts, it may take a bit of detective work to find out what is wrong. The first step is to recognize and appreciate the appearance of a healthy horse, to understand each horse's personality and peculiarities, and to be alerted when the horse deviates from his own particular form of normal behavior.

In general, a healthy horse has a good coat that shines in summer and winter. He has a smooth and elastic skin, and his gums and eyelids are a pinkish color. He's alert and bright-eyed, interested in life around him. When he has the proper amount of flesh and muscle covering his skeletal structure, he is said to be "in good flesh." Once a good feeding schedule has been established for him, he should show eager interest in sticking to it. Some horses, especially nervous ones, are "shy feeders," that is, they always pick at their feed, stall-walk a bit while they chew, eat another bite, and repeat the process. But if they do it all the time, that's normal for them. A healthy horse's manure is bulky and consistent, with a normal smell—rather than watery and foul-smelling.

## Vital Statistics

It is important to know your horse's vital statistics. Check them when you know he's healthy, find out what is normal for him,

---

◄Keeping your horse healthy requires the periodic attention of a good veterinarian, as well as your own regular attention.

and write them down. When you suspect trouble, you can check to see if there is any deviation.

**TEMPERATURE.** The normal body temperature for a mature horse at rest is about 100°F. to 101°F. To check his temperature, you need a good equine rectal thermometer with a string attached to one end of it. (It is made with a handy loop for this purpose.) Leave it in for three minutes to get an accurate reading.

**RESPIRATION.** The simplest way to check your horse's respiration is to place your hand close enough to his nostril to feel him exhale. He should breathe between 8 and 12 times per minute at rest. During normal breathing his nostrils and rib cage move slightly, rather than the flaring nostrils and heavy ribs of labored breathing.

**PULSE.** Place your fingers under your horse's jawbone and find his pulse. While he's resting and comfortable, it should be in the range of 35 to 40 beats per minute, and each beat should resemble the last in strength, evenness, and rhythm.

**CAPILLARY REFILL TIME.** The capillary refill time test is a simple way to gauge your horse's cardiac output. To check his capillary refill time, place your thumb on your horse's gum with enough pressure to cause a white spot. When you remove your thumb, the color should return to normal within one to two seconds. A slowdown in regaining normal color may indicate low blood pressure, dehydration, or shock. In times of emergency, this test can provide valuable information to you and your vet.

# WHEN HE ISN'T WELL

A good owner is sensitive and familiar enough with his horse to know when he deviates from normal, to notice a change in his posture or attitude. His vital statistics may or may not deviate from normal, depending on how sick he is and the nature of his illness. He may simply wrinkle his nostrils and back away from his feed. He may just seem distracted, get a runny nose, or drop his condition. But if the owner is aware, the signs are there to read.

## Colic

Colic is, simply put, a stomach pain or bellyache. It can have numerous causes. It ranges in severity from a gassy stomach, or "pizza belly," which is usually short-lived, to a twisted or malpositioned gut, which may lead to shock and a prolonged, painful death. Colic is the number one killer of horses, and as such must always be taken seriously. The most unfortunate thing about it is that it is usually a result of poor management of the horse.

**WHAT CAUSES IT.** About 85 percent of colic cases result from a gut overcrowded with and damaged by parasites. (See page 76 for more information on parasites.) Colic is also caused by the horse eating moldy hay, uncured grain, or fresh lawn clippings which quickly ferment in his belly; ingesting sand from eating hay, grain, or grass from sandy soil; or finding his way to the grain supply and overeating. Overwatering a hot, tired horse can cause colic, or the horse could have a twisted or malpositioned gut.

**WHAT TO LOOK FOR.** In mild cases the horse may be down for a while, get up again, stall-walk, paw the ground, and try to find a comfortable position. With more intense pain, he may sweat and roll. The pain may ease and return. In severe cases, he will sweat profusely and may become violent and self destructive. As he moves into shock, his body temperature will drop, his heart rate will accelerate and become thready, his capillary refill time will slow, and his gums may turn purple. A horse in this condition may require immediate and expensive surgery.

All cases of suspected colic warrant a call to your veterinarian. After you give him several clues, he will decide how much trouble your horse is in and whether you have an emergency or just a situation that bears watching.

**WHAT THE VET NEEDS TO KNOW.** The vet will need an accurate description of the horse's general state: is he alert but uncomfortable, is he rolling, how much is he sweating, does he seem disoriented and self destructive? After that, the vet will want to know:

- when and what he last ate,
- any changes in his stable routine,
- whether his gum and eye membrane color is normal,
- his pulse rate,
- his capillary refill time,
- his temperature (you may not be able to take it, but if it has dropped, his ears will be cold),
- what sounds you hear, if any, when you press your ear to his belly.

After the vet gets a picture of what is going on, he may give you instructions to walk the horse in order to keep him from injuring himself by rolling and to stimulate his bowel. If you are skilled in giving intramuscular injections, the vet may ask you to administer 10 to 20 milligrams of dipyrone. Perhaps he will ask you to give the horse an oral dose of mineral oil. Follow the vet's instructions and keep in touch with him regarding any changes. Sometimes what appears to be a mild colic abruptly changes into an emergency.

## Thrush

Thrush is caused by a fungus, *Spherophorus necrophorus*, which is present in the soil, and which, under certain conditions, invades your horse's foot and eats the horny tissue. You may recognize it by its oozing, gray appearance and extremely foul odor. Some horses are predisposed to thrush because they have narrow or contracted feet and deep clefts in the frogs (the fleshy, V-shaped underpart of the foot)—a good, moist hiding place for the fungus to grow and spread. Thrush is often associated with horses who stand around in dirty, damp stalls in barns where stall cleaning as well as foot cleaning is neglected. In some cases, it may be the result of bad shoeing.

**WHAT TO DO ABOUT IT.** To get rid of thrush, have the horse's foot trimmed and the crevices around the frog carefully pared away in order to remove some of the surface fungus while rendering the seat of infection easier to reach and to medicate. Then, wrap a hoof pick with cotton and with a good anti-thrush medication (your vet or farrier will recommend one), reach deep into the

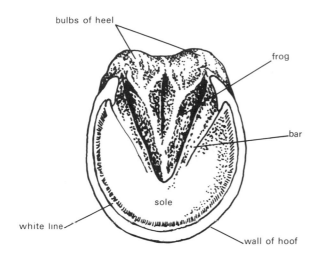

The frog is the fleshy, V-shaped underpart of a horse's foot.

infected crevices and carefully and thoroughly swab them out. Be careful to reach every surface with a liberal dose of medication. Three days of scrupulous treatment should eliminate the problem.

**HOW TO PREVENT IT.** Some horses may pick up thrush in spite of your best efforts, but with a good routine of daily care, you can spot it and stop it before it progresses too far. Keep stalls clean and dry. Use a hoof pick daily to keep your horse's feet clean. Provide plenty of exercise to insure good circulation, and ask your farrier to visit regularly.

## Lameness

Sad to say, unsound horses are often treated as sound by owners who have no idea that they are lame. Not only does this lead to cruelty, but a lame horse forced into work can be extremely dangerous, since exercise can worsen the injury and lead to a bad fall. It need not be that way, for recognizing lameness is not so mysterious. Pinpointing and diagnosing it is another matter, however, and one that is best left to the vet.

**WHAT IS NORMAL.** The first step in recognizing a lame horse is to know a sound one, and the key to soundness is symmetry of motion and stride. At the walk and trot, the horse should put as much weight down on his left front leg as on his right front leg, step for step. The hind legs carry less weight (therefore, 75 percent of lameness is in the front), but, again, the weight should be divided equally between the two hind legs. There is a symmetry that you can see, but besides that, you can clearly hear it. A sound horse, like a metronome, has perfect rhythm and even timing: click - click - click - click. When he is evenly balancing his weight, front and back, his head and neck movements are also regular and slight.

**WHAT IS ABNORMAL.** Front-end lameness can be seen in the disruption of this symmetry, and the most dramatic alteration is in the movement of the *head*. The horse's head is his center of balance, and a horse in pain wants to redistribute his balance, or weight. As the offending painful leg strikes the ground, he hikes up his head in an effort to defy gravity, a reflex action to escape pain. It is plain to see: the head hikes as the painful leg strikes.

## Bleeding

Cuts and lacerations may cause severe bleeding. The most urgent is arterial bleeding, which is bright scarlet and spurts in rhythm with the heart. Bleeding from a vein is less traumatic and easier to control. It can be recognized by its dull color and more even flow.

**WHAT TO DO.** If the horse is frightened, do what is necessary to calm him in order to get his heart rate back to normal, but do not administer a tranquilizer without permission from your vet, since it could induce shock. In treating heavy bleeding, remember that pressure to the wound is always the first step. If the wound is located on a part of the body that cannot be bandaged, apply firm and even pressure manually with a clean towel or bandage, bearing in mind that the goal is to allow a clot to form. If venous bleeding occurs on the leg, wrap the horse with a pressure bandage. For arterial bleeding, apply a tourniquet 3 to 4 inches from the wound, between the injury and the heart. Tighten it carefully until the bleeding stops. Since the tourniquet is cutting off the blood supply to the injured leg, it must not be left on for more than four minutes at a time. Allow the blood to recirculate to the leg, and then reapply the tourniquet as needed until the bleeding stops. Notify the vet as soon as you can.

**HOW TO PREVENT IT.** Since freak accidents do occur, you may not be able to prevent your horse from getting cut, but you can take some steps to narrow the risks. Consider barbed wire your number one enemy and remove any that may be around. It does not mix well with horses. Be sure that pastures are free from debris, and that all passageways that horses move through, such as gates, stalls, and aisles, are free of sharp edges and protrusions. Trim wooded areas where horses stand of all sharp, dead branches to well above head level. Keep fence lines in good repair, for an errant nail or loose board can do a lot of damage.

## Puncture of the Foot

A common injury among horses is puncture of the foot. If there is a rusty nail lying somewhere around the barn or along a fence line, you can bet your horse will find it sooner or later.

**WHAT TO DO.** Common sense would seem to dictate pulling the nail out immediately, but wait! You've got two problems here. The first problem is to keep the horse from driving the nail in further by walking another step on it; so he has got to be immobilized immediately. Most likely, he will be pleased to cooperate. The second problem is that once the nail is withdrawn from his spongy foot, the hole will close and become immediately invisible, rendering it impossible to disinfect the wound. What you must do is pare a hole around the nail with a sharp knife, so that once the nail is removed, the puncture may be found and treated, and the wound may have a channel from which to drain.

After removing the nail, soak the horse's foot in warm epsom salts for 45 minutes, and then swab the wound with cotton soaked in iodine or peroxide. Bandage the foot to keep it scrupulously clean, and notify the vet. The vet may want to improve the drainage hole and give the horse a tetanus booster, antibiotic, or pain reliever. It may take a radiograph to reveal whether the nail punctured bone.

## Gravel or Abscess

A gravel is an infection of the foot caused by a very small object (a splinter or a piece of gravel) that works its way up into the foot and causes havoc. A gravel is usually characterized by a sudden and severe onset of lameness. The infection that results is extremely painful, hard to find, and, therefore, tedious to cure. The abscess often works its way upward and breaks out at the coronet band.

**WHAT TO DO.** Summon the vet. He will probably remove the shoe and search for signs of abscess. If he can find the infection, he'll open a hole for drainage and administer antibiotics, tetanus, and a pain reliever. Whether or not he finds the seat of infection, you'll have to soak the horse's foot in warm epsom salts for about 45 minutes at least three times a day (the more frequently and longer, the better) to draw out the infection.

## Cracked Heel or "Scratches"

Cracked heel, also known as "scratches," is a type of psoriasis that occurs in the fetlock area, generally in the hind feet of horses who are turned out in wet, muddy, or snowy pastures. The skin over the heel becomes red and tender at first, and later cracks appear which deepen and produce hard, calloused edges. There is intermittent bleeding, and pus exudes from the crevices. Unnoticed and untreated, scratches can cause beds of proud flesh known as "grapes." Besides being unsightly, this scar tissue interferes with the natural movement of the joint.

**WHAT TO DO.** In bad weather, a heavy coating of petroleum jelly over the heel and fetlock area will keep out damaging moisture and allow the skin to stay supple. Keep the fetlock area clean and well trimmed at all times. At the first sign of cracking, peel off the scab and apply an antibacterial cream, such as Furacin.

## Laminitis or Founder

Laminitis is circulatory congestion within the horse's feet (usually both front feet, occasionally all four), which leads to enough swelling to cause separation of the laminae, or the outer and inner layers of the hooves. This is akin to peeling your fingernails away from the underlying flesh. It is potentially crippling in that the pressure can be severe enough for the coffin bone (the bone above the frog) to rotate and protrude through the sole of the foot. It is most common in ponies and overweight horses.

**WHAT CAUSES IT.** There are two types of causes: direct and indirect. Direct causes stem from excessive concussion and pressure to the foot. An unfit horse who gallops on a hard road (concussion), or a lame horse who compensates for his pain by loading his weight on his sound leg (pressure), may develop laminitis. It can be caused indirectly when a hot, exhausted horse drinks an excessive amount of cold water after working, or when a hot horse is left standing

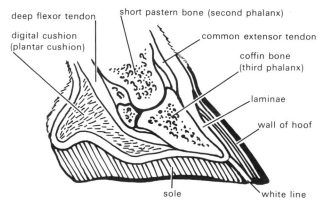

The laminae is the inner layer of the hoof.

around on a chilly day. Laminitis may also occur after a horse ingests excessive amounts of high-protein or rich grass. It may follow an attack of colic, and it may occur if a mare retains the placenta after foaling. Horses who are bedded down on black walnut sawdust have a very high incidence of laminitis.

**HOW YOU RECOGNIZE IT.** A horse with laminitis will be in obvious distress, with an elevated temperature, depression, and an unwillingness to move. His feet will feel hot, and if the condition is in the front feet only, he will stand with his hind feet well under him in order to take pressure from his front feet.

**WHAT TO DO.** When you suspect laminitis, stand the horse in cold water or soft, cold mud and call the vet immediately. Treatment will center around reducing vascular congestion, so the vet will consider various measures such as diuretics and anti-inflammatory agents. Cautious and limited exercise may stimulate circulation and help reduce inflammation.

**PREVENTION.** Prevention methods include proper trimming and shoeing, control of diet, conditioning, and appropriate care following hard work, foaling, and attacks of colic. It is also essential to safeguard horses from being bedded down on black walnut sawdust.

Photo by Margaret Thomas

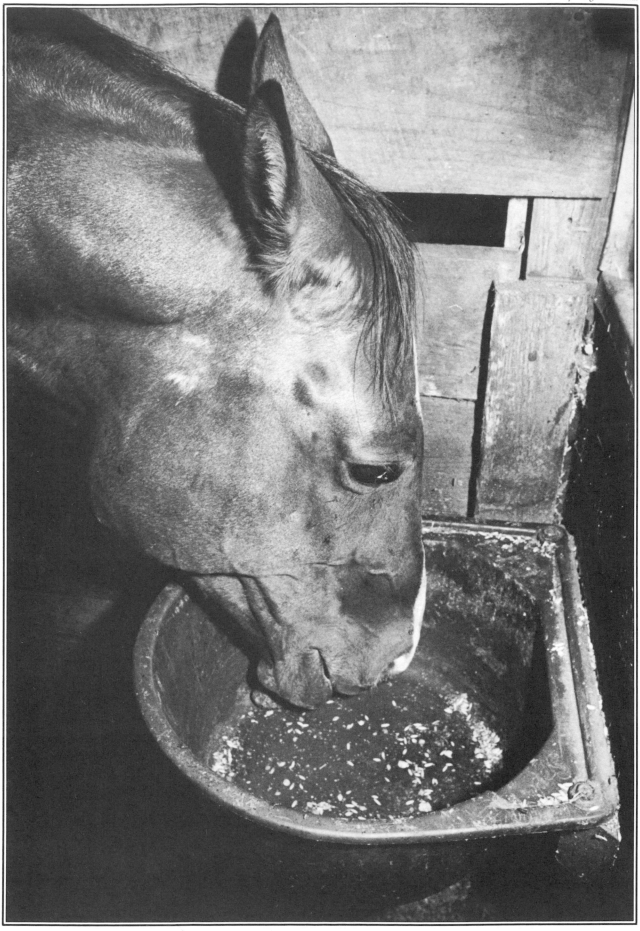

# *Feed*

The horse is a herbivore, a grazer. Left to his own devices, he grazes during the majority of his waking hours, digesting slowly at the same rate that he eats. His digestive system operates best when his stomach is about two-thirds full. When he overloads his stomach, his digestive enzymes don't work efficiently and his intestinal load becomes too heavy, pressing on his internal organs. When an owner takes on the responsibility of providing a horse with his feed, he needs to do it in a way that takes advantage of the horse's preestablished digestive system.

## Maintenance Energy

Energy is the ability to perform work, and it is measured in calories, Btus, or Total Digestible Nutrients (TDN). Maintenance energy is the amount of energy required to sustain life without losing weight. A horse with energy at maintenance level spends most of his time moving from one grassy spot to another, lowering his head to drink, and chewing. Many family horses, idle in winter and worked sporadically in the summer, fall into this category. They are in a maintenance state.

Many factors influence what a horse needs to stay in good condition at maintenance level. His breed is one factor. For example, most Thoroughbred horses have a "big motor" and will usually burn more fuel than a Quarter Horse of the same size. Many ponies require very small amounts of hay and no grain. Temperament is another factor. Climate, age, and state of health all influence the amount of feed a horse needs at maintenance level.

The mature family horse who bears a reasonable resemblance to a genetically sound specimen ought to do perfectly well on plenty of fresh water and roughages containing about 10 percent protein. Some will need concentrates as well. Whatever feeds you choose to give your horse, it is important that they be fresh and clean, free of dust and mold.

The two main categories of feed for horses (besides water) are roughages and concentrates. Roughages are the various types of hays and grasses that are high in fiber and provide your horse with the bulk essential to his digestion. Horses can survive on roughages and water.

Concentrates, as the name implies, are

◄A healthy and happy horse will clean up his feed—it's just a matter of knowing how much to give him.

much higher in food value, or TDN. Oats and corn are concentrates. A horse cannot survive on concentrates and water alone because his digestive system requires the bulk of roughages to function.

# HAYS

There are two basic types of hay: legume hay and grass hay. Legume hays are alfalfa, clover, birdsfoot, trefoil, and lespcdeza. They contain more vitamins, calcium, and protein than grass hay, which makes them more expensive than grass hay. In fact, legume hays provide more protein than the average mature horse needs to maintain good condition. Therefore, most horse owners prefer to feed horses grass hays, such as timothy, Bermuda, bluegrass, fescue, orchard grass, and reed canary, to which they add a little legume hay if the horse needs more protein because of harder work. In most parts of the country, timothy hay is the choice of horse owners, who like to mix it with a clover. However, good grass hay covers the nutritional needs of mature horses at maintenance level.

For a family horse who is doing light or medium work, good hay is the most important ingredient in the feed ration, and in the long run, good quality hay saves you money. You may buy inferior quality for less money, to be sure, but to make up for lost TDN, you'll have to feed more of it, and if your horse doesn't relish it, he won't eat it. The best hay is always the best bargain.

Good hay is soft and springy with an abundance of leaves. It should smell fresh and is generally a green color, although it will bleach to a pale yellow if exposed to sunlight and still maintain its value. By all means avoid hay that has gray moldy patches or seems dusty. Good hay has a minimum of weeds—they just rob your horse of nutrition and reappear in your pasture, through his droppings, next year.

The single most important factor that influences hay is the state of maturity at which it was harvested. You want your hay

## NUTRIENT NEEDS OF HORSES

| water | salt | roughage | concentrates | protein supplement | vitamin supplement |
|-------|------|----------|--------------|--------------------|--------------------|
| maintenance level— mature, idle horse, one hour work/day | | | | | |
| light to medium work— mature horse, 3–5 hours work/day | | | | | |
| broodmares, breeding stallions, yearlings | | | | | |
| lactating mares, foals, weanlings, sick and stressed horses | | | | | |

to have been cut before the legumes flowered or the seed heads opened on the grasses. Other factors that influence the crop are the climate and the richness of the soil in which it was grown. Quality hay comes from serious farmers in good climatic areas. In most of these areas, hay is harvested three times during the season. Generally speaking, the second cutting is preferable.

## Where To Get It

Your county extension agent may help you find good hay—and sometimes if it's a good season and supply is plentiful, local horse owners may divulge their secrets of getting it. Stop in and talk with local farmers. You can look their hay fields over and perhaps make a deal to pick it up in the field, thus saving delivery cost. In some cases, farmers who have ample storage facilities may allow you to contract for a certain amount, which you may pick up as needed.

If you don't live in a good hay-producing area and have to have it shipped in, sight unseen, try to locate a reputable dealer. If you have only one horse, you'll have to get a group of people together who can divvy up a hay order, but that's usually a good way to solve a community problem in an area that doesn't produce its own hay.

If you have to have hay shipped to you by a dealer, you'll have to pay more for it. That may cause you to consider whether you could do better financially by growing your own. However, most agricultural specialists agree that unless you have a ready-made setup for hay production, you probably can't beat commercial costs. Growing your own requires good land with rich soil (fenced to keep stock off), cheap labor, a lot of time, and some extremely expensive equipment which may break down at harvest time, causing crop loss and costly repairs.

For most people who are not farming as a profession, the most cost-effective way to feed your horse is through a commercial dealer.

## How To Feed Hay

The natural position of a horse eating is with his head down, so many people put hay on the floor near the water bucket. This system works fine unless your horse is a stall-walker. In the event that he is, he'll distribute your precious hay throughout his bedding. Therefore, you may prefer a hay rack or a hay net. Whichever one you use, be sure it's high enough so that the horse can't catch a foot in it. People who object to hay in racks or nets point out that a horse is likely to get hay seeds and dust in his eyes. It certainly happens. Do what you think works best for your horse.

However, if he is being fed outside, and you live in an area where the soil is sandy, do not feed him on the ground. The ingestion of sand can cause intestinal blockage and colic. Let your horse eat his hay from a manger, rack, or net.

If your horse is eating hay and grain, give him his hay first. First of all, it will keep him from bolting down his (preferred) grain, because he won't be as hungry when he receives it. And once the gastric juices start activating on the hay, they are ready to go to work on the grain. You'll waste less feed and have a happier horse. (If you ride every morning, give him the larger portion of his feed in the evening. That way, he'll work on a lighter belly.)

## CONCENTRATES

Concentrates are various grains; some of the most common concentrates are oats, corn, barley, bran, and "sweet feed." As your horse's condition and work load warrant, you may want to feed him concentrates. But as you supply more energy from the concentrates, the amount of hay he receives should be reduced.

*Oats* are so commonly, safely, and widely used that when we think of feeding horses, we think of oats. They are very palatable, and their hulls provide roughage and bulk needed in digestion. Oats may be purchased whole or processed through a roller to crack their hulls, which allows them to be masticated more easily, especially by older horses. Oats fed in conjunction with timothy hay is a good ration for mature working horses.

*Corn* should be absolutely dry and about a year old before feeding it to horses. It may be fed on the cob (which takes up more storage space) or shelled. Corn is high in carbohydrates but low in protein. Feeding a quart of corn gives twice the energy as a quart of oats.

*Barley* supplies more TDN than oats and is higher in protein and fats, making it a good choice of feed for the undernourished horse. It's palatable and digestible, but it does have a very hard hull and is usually fed crushed or rolled. It's a heavy feed, which

---

►Age, breed, health, and temperament are just some of the factors that determine a horse's need for the higher food value of concentrates.

some horse owners prefer to mix with bran or oats.

*Bran* is the coarse outercoat of the wheat kernel and has a higher protein content than corn. Horses like bran, and it has a mild laxative effect. It is usually mixed with other feeds and should not exceed more than 15 percent of the concentrated ration.

*Sweet feed* is a mixture of oats, corn, bran, and pellets in a molasses base. It comes complete with a vitamin and mineral supplement of various proportions. You can order sweet feed with protein contents of 10, 12, 14, and 16 percent. For the family horse, 10 percent should suffice.

*Pellets* are also considered concentrates. They are a manufactured feed made of dehydrated hay with vitamins and minerals added. Since they are dehydrated, they use more water from the horse's system, so horses tend to drink more water when fed pellets. Pellets are manufactured for horses and cows, and those for cows are larger and can cause digestive disturbances for horses. Be sure to make it clear to the dealer which kind you want.

## WATER

Horses need a plentiful supply of fresh, clean water. A horse's body weight is about half water. A horse stores 18-20 gallons in his intestinal tract to aid in digestion and to be absorbed into his blood, to carry antibodies throughout his system and waste products out of his system. Water helps

regulate a horse's temperature; without it, his system goes quickly into shock.

On an average day, a 1,000-pound horse will drink 10-12 gallons of water. If he is eating dry feed, such as hay and a pelleted feed, he will need more, especially during hot weather. If he is on lush spring pastures where the water content is high, he will need less.

Even though his system is dependent on water, the horse is finicky and choosy. Horses don't care to drink water that isn't clear and fresh. When he's in his stall, he eats his hay, causing salivation, and frequently sips from his water bucket, mixing saliva and hay with his water supply. The hay quickly ferments at the bottom of his water bucket, and the saliva makes the water scummy. For this reason, buckets should be rinsed and replaced at least once daily.

Horse owners disagree as to the best way to water a stabled horse. Some believe that if a horse drinks water immediately after eating his grain, he will flush the nutrients from his system before they are absorbed. Others feel that free choice of water is the best and most natural system, and I subscribe to that theory. My horses have always done well with this free choice system.

## SALT

The easiest and cheapest way to supply the minerals your horse needs is with a trace mineral salt block. Horses should have free access to it. Salt blocks come in 50-pound blocks for the pasture or bricks for the feed tub. This salt costs only pennies more than plain salt, and it's worth it, for it replaces the minerals horses lose through sweating.

## SUPPLEMENTS

Various methods can be used to analyze the nutritional status of your horse, including hair follicle analysis and blood tests. But they are often costly and not always accurate. The most sensible method is to use your own eyeball. How's his weight? Does his coat shine? Does his muscle tone reflect the kind of life he leads and his breeding?

The fact is that equine nutritionists have not determined exactly what the vitamin and mineral requirements of horses are, and meanwhile, the commercial horse-feed industry pushes expensive supplements—highly concentrated forms of vitamins and minerals usually containing at least 20 percent protein. Many horse owners are manipulated into believing that if they don't feed supplements they are neglecting their horses. In most cases by far, the mature horse who is fed quality hay and concentrates requires no supplements.

There are some exceptions, however. Sick horses benefit from supplements as do broodmares and lactating mares, breeding stallions, young, growing horses, and highly stressed horses, such as those on the Equestrian Team. If you've got a family horse and you're riding an hour or so a day, a 10 percent protein feed should cover his nutritional needs. If he loses a few pounds, just give him more of it. 🐎

# Pasture Management

Because the horse is a grazer, one of the most important things we can do for him is to provide him with good pasture that will make a safe and healthy environment for him and allow him to exercise his herd instinct as well as his body. Young horses in the growing stage mature into healthier specimens if they are brought up on spongy turf rather than in the confinement of stalls with hard floors. Broodmares and idle, mature horses on pasture are better off mentally and physically, and working horses relieve tension and stress by having pasture time. But there is more to it than putting a fence around a field, turning the horses out, and letting nature take its course.

Pastures have a life of their own. They may seem hospitable, but they're tricky. They invite horses to graze while harboring deadly parasites, and they may produce conditions in which weeds and deadly plants thrive. They send up lush green grasses in the spring and then withdraw vitamins and minerals from them. Pastures tend to deteriorate. It is their nature, and they can't be maintained without good management. The climate and the type and fertility of the soil have great influence on the success of a pasture and the nutritional quality of the grass. To make the best of whatever land you have, it's necessary to understand the plant cycle and be familiar with the grasses.

## The Plant Cycle

The growing season for most pastures can be divided into three time periods: from first growth in the spring until June; from July to August; and from September until frost. At the start of the growing season, grasses and legumes send tender leaves toward the sun, which coaxes vitamins, minerals, and proteins into the plants. These tender shoots are the chateaubriand of equine grazing, and they make for well-contented and nutritionally sound horses. The trouble is, this quality is short-lived, for once the plants are so enriched, they gear themselves up for reproduction and send up a seed head or a flower. As the seeds fall, or as the flower is produced, the plant completes its job of preparing for reproduction and turns lazy. Its nutritional value

---

◄ Like horses, pastures require a certain amount of care and attention to function as an asset instead of a liability.

sharply declines, and the leaves turn tough and fibrous. Here is where you step in with the intention of frustrating the plant.

Because pasture grasses are seasonal, it is important to get in as much good grazing as possible over the longest period of time, while maintaining high feed value. This is achieved in part by mowing pastures just before the seed heads and flowers form, which frustrates and rejuvenates the plant. The plant responds by gathering up nutrients and trying again to reproduce, beginning by sending up those tender, nourishing shoots. The horse eats the succulent grass, and you sharpen your mower for the next time around (while saving on your feed bill). To keep pastures at their most nutritious state, clip the grass at least three times during the growing season.

Additional benefits of mowing are control of weeds and brush, and elimination of uneaten clumps of grass and coarse growth caused by incomplete grazing.

## Types of Grasses

Five species of grasses account for most of the seeded pastures in the United States: orchard grass, reed canary, smooth brome-grass, Bermuda grass, and fescue. These grasses are all good, but fescue poses a problem specifically for broodmares.

FESCUE AND BROODMARES. Fescue rates as one of the hardiest and most valuable grasses in the United States. It's a cool-season grass that extends the grazing season; it forms a dense sod which helps to resist erosion and withstand heavy traffic, whether by stock or machines; and it's nourishing, too. But broodmares grazing entirely on fescue have been known to develop a variety of problems. Some mares

deliver a live foal but don't lactate. Some abort in the third trimester; others carry a foal for a prolonged period, up to 14 months, and die trying to deliver an enormous foal. Some foals are delivered in an abnormally thick placenta from which they cannot escape, and they suffocate.

There are steps you can take to reduce the risks, because the missing dietary element that is necessary for the successful production of foals is available in legumes. Seed your fescue with plenty of ladino or red clover; they grow well together. Remove pregnant mares from fescue pastures at least 30 days, and preferably 90 days, prior to foaling, and feed them alfalfa hay. Be sure to keep in touch with your vet during this entire process.

## Pasture Rotation

In most parts of the country one horse can manage on 2½ acres, although some will require supplemental feeding. Owners who have less than 2½ acres per horse would do well to practice pasture rotation to prevent overgrazing and abundant parasites. The smaller the acreage, the more valuable pasture rotation becomes. You can use electric fencing to divide the space in half. Graze one side for two weeks while the other half rests and grows; then switch. The droppings should be picked up in the unused pasture, or at least scattered to expose parasites to sunlight, which kills them. Clearing the pasture of droppings also helps to reduce uneven grazing.

## Combined Grazing

Combined grazing is an efficient management practice; various animals clip grass to different heights, and what one leaves,

▶There are advantages to allowing cows (or other compatible animals) to graze in the same pasture with your horse.

another will benefit from. Horses won't graze close to their own droppings because the grass takes on a bitter taste. (They graze close enough, however, to pick up parasites from the manure.) But cows will graze over horse droppings, with the added advantage that they are not vulnerable to equine parasites.

## Parasites

Parasites are a constant fact of life in pasture management, a threat that must be understood, taken seriously, and defended against. To fight them, use a chain harrow, a rake, or any other device to disturb and redistribute the droppings in the field, exposing the parasites to sunlight and thus killing them. It is a chore you must not avoid. The more crowded the pasture, the more often it is necessary. In expansive, uncrowded pastures, droppings should be rolled at least four times a year, and in small

spaces where rotation is the practice, it should be done every two weeks. This program must be coupled with a good worming program, which your vet will prescribe.

## Poisonous Plants

Most poisonous plants must be eaten in amounts of 1-3 percent of body weight to cause toxicity, and some plants are poisonous only during a certain period of development, at a particular time of the year. Generally speaking, most poisonous plants are not very palatable, and horses with good pastures will usually avoid them. However, where good grazing is scarce, hungry horses may eat plants that could harm them.

Yew trees and shrubs are deadly. Never plant them around the barn. The cherry tree's wilting leaves are highly toxic and so is oleander, that sweet-smelling decorative shrub. Other than these commonly planted

backyard shrubs, plants to destroy in your pasture include: California yellow star thistle (horses like it), ragwort, jimsonweed, foxtail, and castor bean.

The signs most associated with poisoning are diarrhea, colic, convulsions, abnormal gait, and abnormal heart rate, all of which may also be a result of other problems. Poisonous plants are most likely to be found along fence lines and around springs, or in your backyard.

# Behavior of the Horse at Pasture

The subject of horse pastures would be incomplete without mentioning the behavior of horses when they are turned out together. Although at first glance they seem to spend all their time grazing and switching their tails at flies, there is a lot more going on. They communicate by eye contact, nickering, and whinnying. Strangers greet each other by taking turns blowing up one another's nostrils and by following a ritual of squealing and foot stamping. There are probably many other ways they communicate that are too subtle for us to read, but one thing is for sure: they have a definite pecking order.

Dark-colored horses are generally dominant and are the group leaders and enforcers. They tend to be territorial and opinionated, and mares usually have more clout than geldings. The older the horse, the more secure is his place near the top, especially among mares. Large horses usually dominate smaller horses.

When a new horse is introduced to an established group, he will be challenged and separated from the rest of the horses by the dominant horse and his best buddy. The group posture that keeps the new horse separate will usually go on for a week or two, after which the dominant horse, having made his point, will lose interest and everyone else will follow suit.

However, if the new horse should offer an opinion as to who should be boss, the game may take on a vicious tone. Several horses may trap the newcomer in a corner by a fence and kick the dickens out of him. Body blows don't usually do permanent damage, but kicks on the legs tear flesh and break bones.

If at all possible, introduce the newcomer slowly. Either put him in a separate pasture in view of the group and let them get acquainted while separated by a sturdy fence, or put them in the barn together so they can work it out with a stall partition between them. If neither of these options is possible because of your physical setup, pull off the hind shoes of all the horses to lessen the severity of a kick.

## Run-In Sheds

Recently, I heard someone say, "Horses are so dumb they haven't enough sense to come in out of the rain." What the speaker apparently didn't know is that horses at pasture are very well equipped to handle the weather. They may head for the woods or a sheltering hillside when the wind is especially ruthless or when the flies are really bad; but if you put up a run-in shed for them, they'll probably disappoint you as to how often they will use it. Even so, sheds are usually worth the effort, especially if your pasture offers no natural protection. The added advantage is that on the first

---

► A run-in shed can be built with one, two, or three sides closed in. This one includes storage space on the left side.

cold, rainy night after you've built it, when your horse is out grazing, you can fall asleep with a clear conscience.

To protect your horse from the weather, a run-in shed has a back wall and a roof at least 8 feet high. One or both ends may be closed in for better protection or to include a storage area where you can keep hay. The shed should be at least 12 feet deep and 12 feet long for one horse. For every additional horse, another 12 feet should be added to the length. If you can afford to make it even longer, do so, for one dominant horse may declare the community structure his own and fuss at any other horse who attempts to use it. For easy escape, they need the front left open.

## Dogs and Horses

Another problem that can occur in the pasture is horses being chased. Luke the Drifter was a Border collie who wandered onto our property and decided to stay. He liked to go to the pasture behind the barn and kind of bunch the horses up, slowly, carefully, and tactfully. One morning while he was crowding them I gave a hand signal and waved him toward me. I found out that a Border collie takes any hand signal very seriously (just as a horse takes being chased very seriously). Luke took my wave as a sign to "herd." He galloped the horses down through the back gate of the paddock and out the front gate, gathering speed and confidence every step of the way. As they disappeared over the hill in the driveway, dust settled on the sound of his barking. The horses finally parted ways where the gravel road meets the hard top, hoofprints in all directions. By that afternoon when they were back in the pasture, Luke was out helping a neighbor with his horses.

Once a dog starts chasing, it's pretty hard to stop him. A well-aimed blast of birdshot will make some dogs think twice, but it probably will make a herding dog turn sneaky. Luke would just wait until lights out at night to run them ragged. Herding dogs and horses don't mix.

# *Fencing*

Fences increase the value and improve the appearance of a farm. They can be used to separate stock, rotate pastures, and keep out neighboring stock. (In that respect, "good fences make good neighbors.") When considering the type of fence to buy, there are several things to keep in mind:

- cost of installation
- maintenance
- protection and safety for your horses
- aesthetics

The order of the list depends on your budget and the value of your stock. Horses are amazingly accident prone. If there's a way to get hurt, they'll find it. And if there isn't, they'll manufacture it. Like people, horses think that the grass is greener on the other side of the fence, so they tend to lean and push on their fences while reaching over for that special clump of grass (just like the ignored clump by their feet), and then they get into a tangle with the fence.

And don't think your fence-laying efforts will please them; they will display their appreciation for your hard work by chewing the new fence. In their flight from fright or in high-spirited playfulness, they may run into it, and, occasionally, they'll run through it. In this respect they are unlike cows. Cows will get through fences, but they usually don't get hurt. Their skins are much tougher than horses' skins, and they usually exit slowly, kind of wiggling their way through. If you will be grazing horses and cattle together, fence for the horses.

There are many types of fences that will serve you well. Get the best one your budget will allow and build it as carefully as you can. Whatever you choose, a few rules apply that will make your fence safe and long lasting.

- Most horse owners prefer wooden or concrete fence posts to metal posts. Metal posts have sharp edges that can injure a playful horse.
- Contours in the pasture often require many curves in the fence line, but, where practical, straight fence lines will stay tighter.
- Fencing for horses should be 4–4½ feet tall. It helps with the leaning problem and sets up a psychological barrier that horses tend to respect. (A horse who will jump over a 4-foot fence will probably

◀ An attractive post-and-board fence is one of a variety of fencing possibilities available for your consideration.

also jump over a 5-foot fence. I've had a few horses over the years that I never could keep home although my pastures are fenced with 4½-foot post-and-board.)

## Paddocks

Paddocks, or corrals, should be treated differently from pastures. Paddocks are small fenced areas, usually near the barn, where horses are turned out for an hour or so to play, get exercise, and relieve boredom. Horses who are fit usually enjoy their paddock time and gallop and buck for 15 or 20 minutes before settling down. Because paddocks usually hold active horses, galloping the fence line and kicking their heels, paddock fences need to be strong, smooth on the inside, and without corners. Posts should be set no more than 8 feet apart.

Barbed wire, loose mesh wire (American wire), and electric wire are inappropriate for paddocks. Post-and-rail, post-and-board, or diamond wire mesh are safe and very popular. Because paddocks are small areas with very active horses, they require the best fence you can afford.

## DIAMOND WIRE MESH

If you can afford to fence for ultimate safety, you may want to buy diamond wire mesh. Many of the big breeding farms use it. It's seen so frequently in Kentucky that

►Diamond wire mesh with a board on top makes an excellent fence—strong, safe, long lasting, and reasonably attractive.

some people call it Kentucky wire. It's maintenance-free, the horses can't chew it, and best of all, it's woven small and tight enough to be a trustworthy fence for all types of horses. A top board may be added to make it higher and more visible to the horses.

## Assembly

Diamond wire mesh comes with full assembly directions. It's corrosion-proof and has at least a 15-year life span. It's made of horizontal and diagonally-twisted cables that are spaced about 4 inches apart. Posts must be 2-inch steel pipe or treated wood. Corner posts should be set in concrete because the wire is stretched so tightly; posts should be 8 feet long and very hefty. Line posts should be 7½ feet by 5 inches.

The posts should be set at least 3–4 feet into the ground. Diamond mesh is dense and heavy, so it takes a lot of muscle to stretch it. A double-jack fence stretcher or a tractor with a come-along should make it manageable. The fence should be set 8 to 12 inches off the ground to allow space for cutting the grass and controlling weeds.

As with all wire, it's best to install diamond mesh in the summer. Wire installed and stretched to proper tautness in the warm weather allows for contraction on cold days. Wire fences erected in cold weather tend to expand in the summer to the point of sagging. Most wire meshes have crimps in the wire called tension curves; when the proper amount of tension is placed in the fence, the tension curves will stretch out about halfway.

## WIRE MESH

After the diamond mesh, safety is reduced somewhat, but the budget is eased. There are wire meshes of various sizes and strengths, but page wire and American wire are very common. A few rules of thumb make a better fence.

- A top board strengthens a wire fence, impedes sagging, and makes the fence more visible to the horse. Generally, 1 × 6 oak or 1 × 6 treated pine is used.
- The heavier the wire, the longer it will take to rust.
- The smaller the mesh, the stronger the fence. A weak and sagging fence is an invitation to a horse to get a foot caught.
- Avoid any type of square wire mesh that is slightly smaller than horses' hooves. A kick back into a fence can open the wire just enough to get a foot into it—while the forward pull to free it isn't as strong.
- If you use a wire mesh, be sure to have wire cutters where you can lay your hands on them quickly.

Accidents with wire fences usually occur one of two ways. A wire may slip between the horse's hoof and shoe, causing panic during which he may hurt himself by tearing muscles or ligaments. Or he may get a foot through the mesh, panic, pull back, and really tear himself up. Cuts on the pastern are hard to heal because each time the horse moves he disturbs the healing process. An electric wire, run along the top of the fence, much improves wire meshes by keeping horses at a distance. If you keep an eye out for sagging fences and use a top board and/or a hot wire, risk is reduced and you have an economical fence. In spacious pastures where horses don't challenge fence lines as often, page wire or American wire can be a reasonable solution.

## BARBED WIRE

Barbed wire should never be used to fence any area that you intend to turn horses into. It can viciously wound and cut horses, and they can fight entrapment to the point of inflicting unthinkable harm on themselves.

## ELECTRIC WIRE

Electric wire is inexpensive and easy to install. Like other fencing, it also has advantages and disadvantages. Once a horse has touched it, he won't go back for a while. Electric wire for horses should be strung 30 to 36 inches above the ground, and attached to posts or stakes spaced 15 to 30 feet apart, depending on the lay of the land. The closer the posts, the greater the visibility of the fence to the horses, but no matter how close the posts are, it's always a good idea to tie strips of cloth onto the wire. If the horses don't see the wire, they'll see the cloth. Although one strand of electric wire doesn't pull much, braces at the end posts do make a stronger fence.

A bucket or wooden box can be used to provide shelter from the weather for the electric charge. Some fence chargers produce a continuous current, while others, which are safer, produce an electrical impulse for only 1/10 to 3/10 of each second.

Some chargers offer other desirable features, such as a signal light to indicate a short circuit and a switch to beef up power during dry spells. These features are usually worth the extra cost, especially for a permanent electric fence.

There are two problems with electric fence. If it's touched by anything, commonly weeds or a branch, its power is either drained or grounded out. Stretching hot wire through a wooded area would require constant checking. So it's not a fence you can put up, trust, and forget about. Returning from a weekend trip, for instance, you may find your livestock population reduced or absent.

Another problem is that whether it's hot or grounded, it's not a barrier with great visibility that a horse feels respectful toward. A horse can mistakenly or purpose-fully run through it. Breaking through an electric fence probably won't hurt a horse, but it will frighten him, and he may get into trouble once he's loose. And if he breaks through once, you can bet he'll do it again.

The best use of an electric fence is as a temporary barrier for rotating pastures. If you have two horses on four acres, for instance, you might find it useful to divide the area with an electric fence. That way, horses can graze one half while the other half grows. Used properly in small areas and checked frequently, electric fence can be a practical solution to fencing problems.

## WOOD FENCES

Post-and-rail and post-and-board fences are less expensive than diamond mesh, and more expensive than most other wire. They are very popular with horse owners. Though they require some upkeep, they are relatively safe, long lasting, and nice to look at, too.

### Post-and-Board

Post-and-board is slightly more costly than post-and-rail, but it has one advantage. When the boards are nailed to the insides of the posts, the result is a smooth surface that can't catch a horse's knee or hip. Beyond that, nails that come loose are pointed outward and pose less of a threat. Post-and-board is strong enough to hold a horse unless he gallops straight at it, and it's highly visible as well. It can be painted (with lead-free paint) or creosoted to help deter horses from chewing on it.

One of the nice features of this fence is

◄The pipe gate is rounded so that horses can't cut themselves on sharp corners.

that it's versatile. You can nail the boards to the posts in a fashion that solves your particular problem. For the vast majority of family-horse owners, a good and satisfactory fence will be achieved by simply using three boards, one over the other, at a little more than 1-foot intervals, depending on how high you want the fence to be.

Post-and-board should last between 20 and 40 years, depending on the materials and the climate. Frequent wet weather tends to rot the posts; moisture gains entry to the boards by way of the nail holes. Most people cut the tops of the posts at an angle to help water run off, and cover the seams of the boards with a strip of excess board where they are nailed to the fence. This helps repel moisture and keep warping at bay, and it makes a stronger fence as well. Since green boards tend to warp, let them season for a few weeks before nailing them up.

Locust posts and 1 × 6 rough-sawn, 16-foot, oak boards make the strongest fence (and resist chewing). If these woods aren't available, use a pressure-treated soft wood with posts at least 4 inches in diameter. Posts should be 7½ feet long, tamped into holes 3 feet deep. Use three galvanized screw nails on each board.

## Post-and-Rail

An excellent alternative to post-and-board is post-and-rail. Where it is available, it's popular because of its safety and beauty. It's a little less costly than post-and-board and has the same life span and several very nice features. It's assembled without nails and has no narrow spaces or sharp edges for

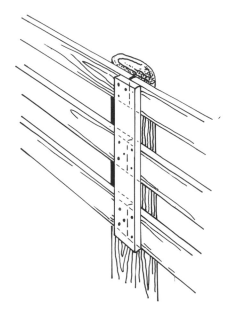

A strip of board to cover the seams helps prevent moisture from getting into the nail holes in the posts.

trapping a hoof. If a rail should break, just supply a new one. A post-and-rail fence doesn't need to be painted and looks attractive as it weathers. If horses do decide to chew on it, a strand of hot wire across the top is inexpensive and easy to install.

Posts come with three or four holes, lending flexibility to your design. If you want a three-rail fence, the post should be 7 feet long. For four rails, use an 8-foot post. When ordering rails, bear in mind that hard woods discourage chewers.

Good posts, well anchored, make good fences, but they are expensive in some areas and scarce in others. Some people prefer to cut and treat posts from their own stand of timber, thereby saving money for other projects. Your local sawmill can probably tell you how to do this. 🐎

# Vices and Bad Habits

There are many reasons that horses develop bad habits. Some are the result of bad management systems, such as long-term confinement, which produces boredom or loneliness in a horse. Some horses develop bad habits simply because they are spoiled or high-strung. Others are the result of fear or an aggressive temperament. Here is a list of the most common vices and what you may be able to do about them.

## WEAVING

A horse is weaving when he stands in one place in his stall and shifts his weight back and forth from side to side for long periods of time. It's a mark of nervousness or boredom. A weaver is an unhappy horse and probably a thin one who worries his weight away. His weaving can get other horses in the barn started, too, because he drives them crazy. He'll dig a hole in the stall floor, which will put undue stress on his legs. Horses who are weavers must be declared weavers when they are sold.

If you have a horse who is weaving, first of all, be sure he has enough pasture time. Let him express himself as a horse. Provide

him with a friend—try putting a goat in the stall with him. If his stall doesn't have an interesting view, try providing him with one that does. Sometimes a toy will help relieve his worries. Try suspending a plastic milk bottle from a string and let him bat it around. And finally, be sure he has plenty of hay to divert him.

## CRIBBING

A horse is said to be cribbing when he puts his top front teeth on a surface (usually a stall door or a fence), opens his mouth and expands his esophagus, and sucks air, making a sound similar to a burp. It's a nervous habit that one horse can learn from another, and it can cause digestive problems and weight loss. Because cribbing can't happen without the expansion of the esophagus, a properly adjusted crib strap will make it physically impossible for the horse to crib. The strap may be bought at any tack shop and should be adjusted snugly, so that the horse may breathe and swallow comfortably, but is unable to suck air. The law requires that cribbers must be declared so when being sold.

◄ Cribbing is a bad habit which can cause digestive problems and weight loss in a horse.

## CHEWING

Most horses who chew wood (stall doors, fences, etc.) are frustrated and bored, and the cause can usually be attributed to too little pasture time or to being left alone all day in a muddy paddock or a stall with no hay.

Once a horse has developed this habit, it is questionable whether you're going to stop it, but here are a few things to try. Give him more pasture time and provide him with a buddy if he is alone. When he is in the barn, increase the number of times a day he receives hay. The point is to distract him from his boredom and neurotic behavior with something pleasant. If he still insists on chewing, tack a tin cover over his favorite chewing spots in the stall and paint your fences with one of the no-chew commercial products (or you can mix your own at home using motor oil and lots of red pepper). Some people use creosote, but in my experience it doesn't help much. For the safety of your horse, be sure that wherever he likes to chew, he's not ingesting leaded paint.

## BITING

Among horses, especially youngsters, biting is often a form of playful affection and well within the range of normal and expected behavior. But when a horse applies his fun-loving capriciousness to people, he must be disciplined immediately and in no uncertain terms, lest the game become a habit. You can expect nipping and biting from young horses, and most stallions are apt to give it a try—it's a healthy policy never to turn your back on one. Stallions may do it out of fun or out of viciousness, but due to their size, strength, and aggressive temperament, they are more difficult to correct. Biting may also be expected of some horses while the girth is being tightened. In order not to provoke a horse to bite, tighten the girth carefully and respectfully, in several stages.

Probably the most common explanation for biters among family horses is the owner himself. People make biters out of well-mannered horses by feeding them tidbits from the hand. Once a horse learns to expect a treat that you don't have, you're going to get it for sure. A kind word and a pat on the neck are much better rewards for a horse than a lump of sugar.

If you are living with a biter, you need to make an unmistakable and clear decision that you won't stand for it. Keep a stout crop in your back pocket while you work around him. Do not provoke the horse to bite, but as soon as he tries, immediately let him have it on the shoulder. It's the quick response on your part that counts. You probably can't hurt him much, but you sure can scare him. If all else fails, put a muzzle on him or sell him.

## REARING

A horse is rearing when he raises his front end and stands on his hind legs. It's a great stunt for a cowboy movie, but in reality it is very dangerous and must be corrected because a rearing horse may fall over backward and do himself and his rider serious damage.

If you are a timid or inexperienced rider and your horse rears, free your feet from both stirrups and just slip off the left side of the horse. Otherwise, be ready. Put a correctly adjusted standing martingale on the horse and wear spurs. The moment he begins to hesitate or to elevate his forehand (his front end), drive him forward with your legs. A horse who is going forward cannot rear at the same time. However, if he manages to rear, drive the spurs into him in a manner that will cause him to associate rearing with pain. Lean forward and grab his mane or wrap your arms around his neck. If you should lose your balance and fall back, holding onto the reins for support, you will probably pull the horse over backward. *Novice riders should not attempt to correct this problem.*

## BARN SOUR

Since the horse has a natural herd instinct and enjoys the company of his buddies and the security of the barn, he may resent leaving it. If he learns that by putting up a big stink he can stick around his equine friends, you've got a horse who is "barn sour." He behaves that way because he has learned to get away with it, and in that sense he's spoiled.

As with rearing, it takes a pretty good rider to correct a horse with this problem, but it can be done. The rider should wear spurs and carry a crop in one hand. He should drive the horse forward forcefully, keeping his head straight. Generally, when a horse doesn't want to move forward, he'll back up, wheel around, or rear. Be ready, be extremely forceful, and be sure you're not expected home for lunch. Winning an argument with a spoiled, opinionated horse takes some time. If you aren't sure you're ready to try, save yourself some trouble and get a seasoned rider to give you a hand.

## SHYING

Shying may be caused by playfulness, poor eyesight, or real fear. It's especially hard to distinguish between a horse's eyesight and his fear, and either one should be treated respectfully but firmly. The most helpful attitude for a rider to have toward his horse is a desire to build the horse's confidence. A horse who is terrified that he is going to be grabbed by the ghost lurking in that old pile of rocks should not be made to go to the rock pile, step on it, and introduce himself. If he's really scared, calm him. In a soothing voice, explain to him what a dummy he is. If convenient, have someone hold him while he watches you tramp about on the rocks yourself. (Be careful!) Help him to learn to trust you. Make a point of jogging him by the rock pile again the next day, encouraging him as you approach it. Soon enough, he'll take it in stride.

If the horse is shying out of playfulness, however, handle it differently. A sharp smack with a crop should be enough to remind him to watch his manners and be more businesslike.

## KICKING

Some horses are confirmed kickers who will strike at anything or anyone within range. Since a well-placed kick can do terrible damage, a kicker is not a suitable family horse, and by all means do not keep one in the barn. People who ride kickers in the company of other horses are obliged to tie a red ribbon in the horse's tail to serve as a warning to stay out of the way. Most kickers are mares. I know of no way of correcting this problem.

# *Shipping*

Sooner or later, you may want to transport your horse to a show, a fair, or perhaps simply to a new location. There is more than one way to do this, and there are many things to consider before you lead that animal up the ramp. Should I buy or rent a trailer? Should I buy a new trailer or a second-hand one? If I buy a second-hand model, what should I look for? What about hiring a commercial carrier? Is it dangerous to transport a horse? How much insurance coverage should I obtain? And finally, once I'm ready to go, what is the right way to get that 1,000-pound animal aboard?

## CHOOSING A TRAILER

The question of whether to buy or rent a trailer depends on how often you will need to transport your horse. If you plan to spend weekends traveling with your horse to some horse-related event, then it's probably best to bite the bullet and buy one. You've only got to decide what size trailer you need and whether it should be new or used. That decision may be influenced by the kind of vehicle you will use to pull it.

If you are hauling ponies or small horses, the Quarter Horse or Arabian horse-size trailer should fit the bill. They are usually about 6 feet high, 9 feet long, and 5 feet wide.

Some of these smaller makes don't have ramps. Horses must step up into the trailer, and it's a little spooky for some horses to have to jump into a dark, enclosed space. Others don't mind whatsoever. Consider the personality of your horse, and know how well he loads before you decide whether to get a trailer that's a step-up model or one with a ramp.

For larger horses, those over 16 hands, you'll need the Thoroughbred size. It will give you about a foot more space in all directions. Extra-long trailers have storage space for feed and tack and even dressing rooms. But what your horse needs is comfort and safety. He needs to be able to stand comfortably, to brace himself by spreading his legs, and to carry his head in its natural position. Too much space will make it possible for him to turn around and look over the tailgate to see where he's been.

Used trailers will definitely save you money, and many good bargains can be

---

◄There are many reasons that you might want to ship your horse at some point—and a few things to know if you do.

found. Trailer dealers usually have a supply on hand, and tack shops and the classified section of the newspaper are full of ads for used trailers.

When you're shopping for a used trailer, you should look for:

- Four good tires and a mounted spare, at least 4-ply (6-ply is better).
- Sturdy flooring. Check under the mat for signs of rotting. Urine can do untold damage to a trailer floor in a few seasons. The floor is often the first thing to go.
- Good, non-skid matting for floor and ramp.
- A first-rate hitch with safety chains.
- All lights in good working order. This includes brake lights, running lights, and turn-signal lights.
- A sturdy trailer frame.
- Latches on tailgate and escape door in good working order.
- Cross-tie rings and chest bar well-made and sturdy.

States have different requirements for safety inspections of trailers. Be sure you know what is required in your state before buying a used trailer.

# COMMERCIAL TRANSPORTATION

If you decide not to buy a trailer because of the limited number of times you'll use it, you may need a commercial hauler for that rare event. These folks are usually listed in the Yellow Pages under "Horse Transportation." Call a few, ask them for local references, and check up on them. Most commercial carriers do an excellent job, but there are some fly-by-nights who put your horse in jeopardy with bad equipment, bad handlers, and bad drivers. Be sure you know the difference.

For long-distance shipping, the Interstate Commerce Commission has guidelines as to how much you can be charged. But the way to spend the least amount of money is to plan well in advance with the carrier so that he can transport your horse with his van filled. His equipment is expensive to run, and he can ship six, eight, or twelve horses more cheaply than he can ship yours alone.

When planning a long-distance trip—cross country, let's say—remember that the air-cushioned commercial carrier, with its long wheel base, is easier on your horse (although harder on your pocketbook) than a ride in your trailer. Most commercial drivers know horses, and they can handle any problems that may occur. For an extra charge, the company will send along an attendant to ride with your horse, and the vans are equipped with buzzers or an intercom system in case of trouble. Since shipping is hard on horses, the carriers have lay-over barns where they can stable the horses for the night, allowing them to move around and get a good night's rest.

## Insurance

Most commercial haulers carry only a small amount of insurance on their cargo. If you want to insure your horse for what he's worth, you have several choices.

Most owners insure their horses for one year and renew the policy annually. But if you're thinking of insuring him specifically because he's being shipped somewhere, there are three policies offered by most underwriters.

**FULL MORTALITY.** This policy covers your horse in case of a broken leg, a wrecked van, a heart attack, or the like—any internal or external injury from which he should die during the trip. To make a claim, however, a vet certificate is required. This policy covers the horse from the time of loading to unloading.

**ACCIDENT INSURANCE.** This policy covers external problems like impalement, cuts, and bruises. A vet certificate is not required. The horse is insured from the time of loading to unloading.

**THIRTY-DAY POLICY.** This policy covers your horse for thirty days from the start of the trip. It's designed to cover problems that show up after the trip, like shipping fever, pneumonia, or tendon strain. Sometimes horses get sick on a trip but don't display the signs until after unloading.

## PRECAUTIONS

Some horses just don't like to be shipped. High-strung horses, spooky horses, and yearlings traveling for the first time are the most likely candidates to have problems. They fret, worry, and sweat profusely. A wet, sweating horse in cold weather can catch a cold or pneumonia. Some horses are scramblers. When the vehicle is moving, they thrash around; sometimes they even fall in the stall, panic, and injure themselves. Occasionally, a horse will try to jump over the chest bar and get hung up on his hind legs, without enough space to move forward and too unbalanced to move back again. (There have been cases requiring a welder to take a trailer apart to get the horse out.) It's not uncommon for a horse to break his halter on the cross-ties and turn around and try to jump out the back. Finicky horses won't eat or drink during a trip and lose condition. Horses that ship well are much blessed by the shippers and by their owners.

Horses who are known to be poor shippers should be tranquilized. It may save the horse, as well as the trailer, from unnecessary stress and potential injury. Ask your vet what to give him and how often.

For long trips, remove the horse's shoes. If he should step on himself while balancing or scrambling, he won't do as much damage. By all means, whatever distance you're going, put shipping boots on him or bandage him. Bandages will help support his legs on long trips and will protect him from cutting himself.

Provide your own hay and water. That's what your horse is used to, and he may not want to touch a strange supply.

Quick changes in temperature can give a horse respiratory problems or pneumonia. Leaving Florida one January morning and arriving in Minnesota the following night can make the healthiest horse sick. Be aware of how much air he's getting, and take along his blankets. Make the transition from climate to climate as smooth as possible.

You may want to wrap a tail bandage around your horse's tail while he's being shipped. Some horses lean or push back, and the result is that the horse's tail looks more like a rodent's tail at the end of the trip. Wrapping it just protects it from the rubbing.

Horses who don't like shipping and may throw a fit should have a head bumper. It's like a padded crash helmet with ear holes for horses. It will protect his head, in case he rears, from smashing into the steel braces of the vehicle. Your local tack shop can supply you with one.

## LOADING

Some horses object to being loaded, and it takes time and patience to overcome this attitude. If you handle the situation quietly, without showing temper or testing his strength against yours, you'll get him in the trailer sooner or later. Start by parking the trailer so that it catches as much light inside as possible, hang a hay net in front of the stall and open the escape door. Seeing the inside of the trailer clearly and focusing on the hay net may be all it takes. If not, try coaxing him with a bucket of grain. Give him a bite and then take a step back. Speak soothingly to him and give him confidence. No good? You now need two people.

Have a helper place the horse's front feet on the ramp. He may simply need to know that it's sturdy, and that he won't slip. If he won't go in at this point, he'll probably start pulling back hard, so be sure you don't have the lead shank wrapped around your hand. Once he's started this game, counter by putting the shank through the left-side ring of his halter, under his lip and over his gum, and snapping it on the right-side ring. He'll understand that by pulling back he's hurting himself, but he may carry on anyway. At this point, it is well to remember that the horse is thinking about avoiding going into the trailer, rather than your welfare. The helper must never stand behind him.

Still at a standoff? Have the helper get a broom and repeatedly poke the horse, once every few seconds, on the rump. The aim is to make the horse so annoyed at being outside the trailer that he'd rather be in it. If this doesn't do the trick, get another helper and have each of them hold a longe line behind the horse, just over his hocks, and coax him up. If you have to begin at the beginning, do so. Remember, your concentration and self control are better than his. You'll win by tactics, not by strength. When he does go in, have your helper immediately do up the chain behind him. Then tie his head and close the ramp.

When carrying one horse in a two-horse trailer, be sure to load the horse in the left stall. Since roads tend to have a hump in the middle, this technique will help balance the trailer.

Loading problems can often be avoided by giving the horse an opportunity to become familiar with a trailer in advance and to associate it with pleasant moments. To achieve this, park the trailer in the pasture in the direction in which it receives the most light inside. Feed the horse hay from the ramp, so that he must put his feet on it to eat. After a week or two, move the hay further into the interior of the trailer. Finally, when he's completely comfortable with the idea, hang the hay in the hay net and close the trailer. Stay with the horse to comfort him and give him confidence, and pretty soon he will consider his trailer to be the best restaurant in town.

I've known only one horse who could not be loaded. He was owned by a woman who once took him to a horse show. She was dressed immaculately, and her horse and tack were gleaming, but her trailer was dusty. She decided to pull the rig, horse and all, through a car wash. After that experience the horse could never be persuaded by any means to set foot on or near a trailer again, and as far as I'm concerned, the horse made a fair and reasonable decision. 🐎

# An Introduction to Riding

*"The trouble with most horses relates directly
to who is in the saddle."*

This chapter has been devised as a teaching aid to help the rider to visualize and become sensitive to some of the basic principles of riding. I don't pretend to treat the subject in depth, and no one can learn to ride from a book. Riding is doing. If you are a beginner, ask around and find a good, seasoned, and dedicated instructor, one with whom you feel comfortable, and begin lessons. The more lessons you take and the more horses you ride, the sooner you will develop a good sense of horses and horsemanship. There is only one way to ride correctly, and that is in harmony with the horse. A good instructor can present the techniques necessary to achieve this goal while keeping you from developing bad habits which are hard to break.

However, there are distinct advantages to seeing photographs of a good young rider who has mastered the basics, on an attractive, well-schooled horse. The rider pictured in this chapter is named Heidi, and she is 15 years old. She began riding only four years ago, and if she keeps on working as hard as she has been, she could go to the top. The gray mare is named Windsor. She's a Percheron-Thoroughbred cross who is good-natured, athletic, versatile, and sound. Windsor has successfully shown, hunted, and won endurance rides—and she still loves to teach kids to ride.

I hope that the descriptions presented in this chapter, along with the photographs of Heidi and Windsor, will give you an idea of how much goes into riding a horse properly, and will complement whatever instruction you receive.

## MOUNTING

To get on a horse, stand at his left shoulder, facing his hindquarters. Hold the reins evenly, with a slight tension, with the left hand. With the same hand, grab hold of the mane or saddle. With your right hand, place the stirrup iron under the ball of your left foot and swing up. Be sure to avoid banging the horse's back behind the saddle with your right leg. Land gently in the center of the saddle. Remember, the first impression the horse has about you is how you mount. Do it quickly, gently, and respectfully.

*Don't* face the front of the horse (his forehand) as you mount, and don't leave the

---

◄Heidi and Windsor make an attractive and animated pair, and they work well together.

To mount, your left shoulder should be near the horse. Hold the stirrup with your right hand.

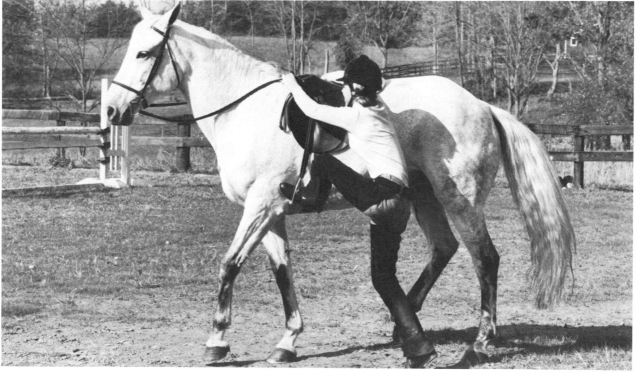

*Don't* face the front of the horse as you mount—you may cause the horse to walk forward, leaving you hopping awkwardly behind him.

reins loose—should he walk forward, you are left hopping behind him. Since he can walk on four legs faster than you can on one, you're out of control, and out of luck.

## THE CORRECT POSITION

Sit lightly in the saddle on your seat bones. The lightness of your seat in the saddle will help keep your hips flexible and your back supple. Your head should be up, with eyes forward, and your back should be straight. Your legs should make contact with the saddle at the inner thighs, knees, and calves. Your ankles should be flexed but supple and relaxed, with the toes turning out slightly, according to your own conformation, or build.

The stirrup should be under the ball of your foot, level with the girth, and your heel well down. Your body weight should be distributed evenly over both sides of the horse. You should feel that your seat is as deep in the saddle as possible and that your lower weight is dropped into your heels as far as possible, while your body above the seat is as light as possible, as though suspended on a string.

Place the reins between your ring finger and your baby finger, and grasp the ends between your index finger and your thumb. Your hands should be gentle but firm. The ends, or "bite," of the rein should fall to the right of the horse's neck. Your elbows should fall slightly in front of your shoulders. A straight line should exist from the horse's mouth to your hand and elbow. This straight line ensures that your hands are in the most effective position to communicate with the horse.

Sit in the middle of the saddle, on your seat bones, head up, eyes front, and back straight.

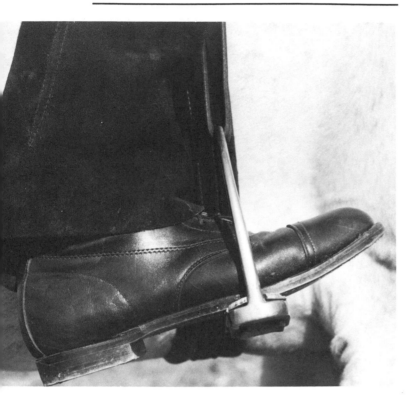

Hold the reins between your ring finger and your baby finger. Grasp the ends between your index finger and your thumb.

# THE WALK

Sitting correctly in the saddle, "close your leg" on the horse, encouraging him to move forward by the pressure of your legs on his sides. As a horse walks, his head flexes forward and back. Your wrists and fingers must flex with the movement of the horse's head to keep a slight, steady pressure, or contact, on his mouth. The speed of a walk is about 4 miles per hour.

## To Turn

The turn requires that you first look in the direction that you want to go. It sounds so simple, yet the act of turning your head affects your whole body. As your head turns to the left, your left shoulder drops slightly, your spine turns, and your seat bone engages against the saddle as your left leg moves toward the horse. Subtle, yes, but the horse feels it. As you turn your head, the fingers of your left hand close slightly on the rein, and a slight pressure is applied to the horse's mouth. When the horse responds, you thank him by releasing the pressure and returning your hand to a soft, flexible attitude.

## To Back Up

To back the horse, close your legs and hands at the same time. (To "close your hands" is to close your fingers on the rein and pull back.) The horse knows he is being asked to move because of the leg pressure, but the pressure on his bridle causes him to move, to yield, away from the bit, and thus he will step backward. He should be immediately rewarded with a relaxed hand.

At a walk, your hands must flex with the movement of the horse's head to maintain a steady contact with his mouth.

## THE TROT

A horse carries his head slightly higher at the trot than he does while walking. Since the slightly elevated head shortens the distance between the horse's mouth and the rider's hand, you must first shorten your reins to maintain contact with the horse's mouth. As you shorten your reins and close your leg on the horse, he will pick up a trot. At the trot (about 8 miles per hour), a horse's head is steady, and therefore your hands are still.

Your body should be inclined a little more forward at the trot than at the walk to maintain your center of gravity over the horse. The trot is a simple rhythm, a two-beat gait. One-two, one-two. "Posting" is your way of keeping your balance, harmony, and rhythm synchronized to the horse. Posting is alternating the rising out of the saddle and the returning to it with each beat. Beat one is rising, beat two returning. Or, look at it this way, beat one is feet, beat two is seat. Up-down, up-down to the rhythm. After some practice, you will find that the impulsion of the horse underneath you actually helps push you out of the saddle, to which you gently return.

## THE CANTER

To ask for the canter, you begin by shortening your reins again, for the horse will elevate his head slightly higher than at the trot. You should sit down in the saddle to make the transition and close your legs on the horse. Your body should be less forward than at the trot; your heels should be down and your ankles flexible, absorbing the shock.

The canter moves at about 13 miles per hour. The engagement of the horse's hindquarters and the impulsion required for this gait make it necessary for your back to be very flexible in order for you to remain deep in the saddle. It is the same principle as bouncing a balloon on a board. Blown up to its capacity and stretched tight, the balloon will bounce higher when hit from underneath by the board than it would if its body had more give, was less tight and more supple. To stay in harmony with your horse, you must absorb his movement in your ankles, seat, shoulders, and especially your back.

## THE GALLOP

The gallop is the fastest gait (18 miles or more per hour). As the horse moves into it, his impulsion increasingly comes from his hind end. He extends his head and neck, thereby moving his center of gravity forward. In order for you to maintain harmony with the horse, you must move your center of gravity forward as well. This is accomplished by shortening the reins, leaning forward, and standing on the stirrup irons.

Your weight should be well down in your heels. You are riding with "two-point contact," that is to say, with your seat out of the saddle. (The two points are your two feet; "three-point contact" includes your seat.) Your elbows should be well in front of your rib cage, and you should maintain a steady feel of the horse's mouth.

## CAVALLETTI

A cavalletti is a series of heavy rails placed not more than 8 inches off the ground and spaced for a horse's trotting stride, which is usually between 4½ and 5 feet long (ponies' will be shorter). Cavalletti provide an excellent suppling exercise for the horse, and they are a good way to prepare a rider to jump. They teach you to control your speed, be aware of your horse's rhythm, keep your head and eyes up, and ride a straight line.

Your hands should be about halfway up the horse's neck, on the crest, allowing him freedom to use his head for balance and to support you at the same time. You must look straight ahead—not down at the ground. Your flexible ankles and deep heel will help absorb the horse's movement as he crosses the cavalletti.

Once you have mastered the cavalletti, you should be able to jump a low fence.

## JUMPING

In thinking about jumping, remember what we said about the horse's skeletal structure. His head and neck are very flexible, and he uses them as a unit to balance himself. His spine behind the withers is surprisingly

Cavalletti are a good way to prepare for jumping. You must keep your head up and your back straight but inclined forward.

Your hands should be about halfway up the horse's neck, your eyes front, and your heels way down.

Coming out of the cavalletti, Heidi is already
in the correct position to jump this low fence.

Over they go in excellent form—it isn't quite as
easy as it looks.

inflexible. During all phases of jumping, you must allow your horse the freedom he needs to use himself properly by keeping your center of gravity over his, and by giving him use of his front end.

There are three phases of the jump. They are take-off, suspension, and landing. At take-off, you release the horse by sliding your hands forward, fingers softly holding the reins. Impulsion from the hindquarters and the upward movement of the horse require you to be forward, your seat just out of the saddle. This puts your center of gravity over the horse and removes your weight from his hind end. Your upper body is supported by your legs, and your legs are just behind the girth, where they are still. Your head should be up, your eyes forward, and your back flat. As always, heels are down and ankles are flexible.

Suspension is the phase when the horse's legs are off the ground and he is airborne. He is stretched out, really using his head and neck, and his legs are tucked up under him. Your seat is close to, but out of, the saddle—head and eyes up, back flat, and heels down. Your hands are supported gently on his crest.

In landing, the horse's head and neck rise up to balance against his downward movement, as his legs and shoulders come down toward the ground. You prepare yourself for a moment of inertia and impact on landing by remembering that your ankle and heel are your shock absorbers and by keeping your weight way down in your heel. Your head must be up to prevent falling forward, and your back should begin to straighten up as your seat returns to the saddle when the horse touches the ground.

The next fence is about 3 feet, 9 inches, and Heidi has asked Windsor to canter down to it.

The jump seems like a good idea to everyone
but Windsor . . .

. . . who quits at the last minute!

. . . and she disciplines Windsor with a sharp
smack behind the saddle.

Instead of turning Windsor away from the
fence, Heidi backs her up, so Windsor can look
at the fence while Heidi reminds her who's boss.

After cantering down to the fence again,
Windsor responds to Heidi's leg pressure by
jumping very well.

Their form is excellent once again . . .

. . . Heidi's head is up, her back flat, and her
heels down.

To dismount, throw your right leg over the saddle, bring your feet together, and jump down to solid ground.

## TO STOP

When you decide to stop the horse, you must return your body to a vertical position and brace your back. This signals the horse that a change is taking place. Close your fingers on the rein and pull back, your hands flexing to the movement of the horse's stride. Pull, release, pull, release. The horse will yield to the pressure on the bridle by ceasing to move forward into it. Immediately reward him by opening your fingers and releasing the pressure on his mouth.

## DISMOUNTING

To dismount, have the horse standing still and square. Hold the reins evenly in your left hand so that if the horse begins to move, you have control. With the reins in the left hand, support your body on the withers of the horse or on the pommel (the front) of the saddle. Disengage your right foot from the stirrup iron and swing your right leg behind the saddle and over the center of the horse's spine, being careful not to bump the horse with your leg. As your leg swings over, hold the cantle (the back) of the saddle with your right hand for support. Still standing on the left iron, bring your right foot to your left foot, ankles together. Disengage your left foot from the stirrup iron, and drop to the ground.

Taking into account that standing on one stirrup, half on or off the horse, may sometimes be a little risky, here is a quick, easy, and safe method for dismounting that you may want to use if you are alone (no one to hold the horse), or if your horse may misbehave. Simply remove both feet from the stirrup irons, hold the reins exactly as described above, and, in one movement, bring your right leg over the horse and slip to the ground without a pause. 🏇

# A Final Word

We've talked about some of the important issues involved in owning a horse: his selection, his structure, his mentality, his health, his habits, and how best to deal with him in various situations. It's been fun putting it to paper, but now that I look back at it, the whole thing can be summed up in four rules:

1. Keep safety in mind at all times.
2. Be considerate of your horse—be a responsible manager.
3. Learn everything you can.
4. Most important of all, have fun!

Photo by Margaret Thomas

# Bibliography

Bone, J.F., D.V.M., et al., eds., *Equine Medicine and Surgery* (Santa Barbara, Calif.: American Veterinary Publications, Inc., 1963).

Brady, Irene, *America's Horses and Ponies* (Boston: Houghton Mifflin Company, 1969).

Codrington, Lt. Col. W.S., M.R.C.V.S., *Know Your Horse* (London: J.A. Allen & Co., 1966).

Conn, Dr. George H., *An Experts' Guide to Horse Selection and Care for Beginners* (Hollywood, Calif.: Wilshire Book Co., 1969).

Dillon, Jane Marshall, *School for Young Riders* (New York: ARCO Publishing Company, Inc., 1958).

Edwards, Elwyn Hartley, ed., *Encyclopedia of the Horse* (London: Octopus Books, Ltd., 1981).

Edwards, Gladys Brown, *Know the Arabian Horse* (Omaha, Neb.: Farnam Horse Library, 1971).

Ensminger, M.E., Ph.D., ed., *Stud Managers Handbook*, Vol. 16 (Clovis, Calif.: Agriservices Foundation, Inc., 1980).

Faudel-Phillips, Mrs. O., F.I.H., *Keeping a Pony at Grass* (Kenilworth, Warwickshire, England: The British Horse Society National Equestrian Center, n.d.).

Froissard, Jean, *Jumping, Learning and Teaching* (Hollywood, Calif.: Wilshire Book Company, 1971).

Green, Carol, *Tack Explained: A Horseman's Handbook* (New York: ARCO Publishing Company, Inc., 1977).

Hanauer, Elsie, *The Science of Equine Feeding* (New York: A.S. Barnes & Co., 1973).

Harper, Frederick, *Top Form Book of Horse Care* (Merck & Co., Inc., 1977).

Hayes, Captain M. Horace, F.R.C.V.S., *Veterinary Notes for Horse Owners* (New York: ARCO Publishing Company, Inc., 1968).

Kelly, J.F., *Dealing With Horses* (New York: ARCO Publishing Company, Inc., 1976).

Lyon, W.E., ed., *First Aid for the Horse Owner* (London: Collins, 1951).

Morris, George H., *Hunter Seat Equitation* (New York: Doubleday & Co., Inc., 1979).

Müseler, W., *Riding Logic* (New York: ARCO Publishing Company, Inc., 1978).

Patten, John W., *The Light Horse Breeds: Their Origin, Characteristics and Principal Uses* (New York: Bonanza Books, 1960).

Pittenger, Peggy Jett, *The Back Yard Horse* (Hollywood, Calif.: Wilshire Book Co., 1964).

Rooney, Dr. James R., *The Lame Horse: Causes, Symptoms, and Treatment* (Hollywood, Calif.: Wilshire Book Co., 1974).

Saunders, George, M.D., *Your Horse: His Selection, Stabling, and Care* (New York: D. Van Nostrand Company, Inc., 1954).

Self, Charles, *Winning Through Grooming* (Omaha, Neb.: Farnam Companies, Inc., 1979).

Self, Margaret Cabell, *How To Buy the Right Horse* (Omaha, Neb.: Farnam Horse Library, 1971).

Self, Margaret Cabell, *The Nature of the Horse* (New York: ARCO Publishing Company, Inc., 1974).

Self, Margaret Cabell, *The Young Rider and His First Pony* (New York: ARCO Publishing Company, Inc., 1969).

Simpson, George Gaylord, *Horses* (New York: Oxford University Press, 1951).

Smythe, R.H., M.R.C.V.S., *The Horse: Structure and Movement* (London: J.A. Allen & Co., Ltd., 1967).

Taylor, Louis, *An Experts' Guide to Horseback Riding for Beginners* (Hollywood, Calif.: Wilshire Book Company, 1978).

Tuke, Diana R., *Feeding Your Horse* (London: J.A. Allen & Co., Ltd., 1980).

Webber, Toni, *Know Your Horses* (New York: Rand McNally & Company, 1977).

Weikel, Bill, ed., *Know the American Quarter Horse* (Omaha, Neb.: Farnam Horse Library, 1971).

\* \* \*

In addition to these sources, numerous volumes of *Equus* and *Practical Horseman* were consulted.

# Index

The numbers in boldface refer to illustrations.

---

# *Other Storey Titles You Will Enjoy*